German Seaplane Fi of WWI

A Centennial Perspective on Great War Seaplanes

Jack Herris

Great War Aviation Centennial Series #2

This book is dedicated to the pioneer Naval Aviators of the first Great War in the air.

Acknowledgements

I want to especially thank Colin Owers for providing many of the contemporary aircraft photos from WWI and for his drawings of the Brandenburg W16 and W25. Martin Digmayer created the rest of the drawings. All drawings are reproduced to 1/48 scale. Thanks also to Bob Pearson for his color profiles and Aaron Weaver for photo editing and cover design. Finally, thanks to Mike Perkins for proof-reading. Any errors are my responsibility.

This is Update 1 published in October 2013; it corrects some typographical errors in the original and adds a recently-discovered photograph of the Lübeck-Travemünde F3 single-seat fighter.

Cover and chapter paintings by Steve Anderson. Please see his website: **www.anderson-art.com**

Color aircraft profiles © Bob Pearson. Purchase his CD of WWI aircraft profiles for $50 US/Canadian, 40 €, or £30, airmail postage included, via Paypal to Bob at: **bpearson@kaien.net**

For our aviation books in print and electronic format, please see our website at: **www.aeronautbooks.com**. You may contact me at **jherris@verizon.net**.

An edition of this book designed for the iPad is available in the iTunes store; search ISBN **978-1-935881-51-3**

Interested in WWI aviation? Join The League of WWI Aviation Historians (**www.overthefront.com**) and Cross & Cockade International (**www.crossandcockade.com**).

www.aeronautbooks.com

Above: The logo of the German Axial propeller company.

Above: The logo of the German Garuda propeller company.

Below: The logo of the German Behrend & Ruggebrecht propeller company on a Behrend & Ruggebrecht propeller.

Table of Contents

Above: *Lt. d.RMI* Fritz Hammer, flying the KDW prototype, Marine Number 748, from the German naval air station at Angernsee, downs a Russian four-engine Sikorski Il'ya Mouromets reconnaissance-bomber on 23 September 1916. Sikorski *IM-6* crash-landed at its base with 293 bullet holes and three of its four crewmen wounded. This was one of only three air-to-air victories scored over these tough bombers during the war.

Introduction

The birth of the airplane had an impact on naval operations in WWI similar to that on land warfare. Like the generals, the admirals immediately saw the possibilitics of naval aviation for long range scouting and direction of naval gunfire. However, the question was whether the new technology was mature enough to be practical.

Naval aviation means flight over water, and that puts a premium on reliability. Many early aviators were forced down by a minor ignition or fuel system fault they were able to correct after landing, then were able to resume their mission. However, faults over water meant the aircraft had to land on water. If a landplane has to ditch, the flight is over and the crew are in grave peril. So naval aviation focused on floatplanes and flying boats to improve the odds of successful mission completion and crew survival. Providing the sea is not too rough, the seaplane or flying boat may be able to float indefinitely, or even take off again if repair is possible, so structural robustness is especially important for naval aircraft; the additional airframe strength required plus the floats or boat hull make seaplanes heavier than landplanes.

Considering that the best qualities of airships, long range, long endurance, and relatively good reliability, closely matched the critical requirements of naval aviation, it is not surprising that airships played a major role in WWI naval aviation. In Germany, which led the world in airship technology, Zeppelins of the Naval Airship Division found more employment for North Sea scouting that any other mission, including their highly publicized raids over Britain. When the German Army became disenchanted with the cost and vulnerability of airships, their airships were transferred to the German Navy. In fact, long-range, long-endurance scouting missions by Zeppelins over the North Sea were essential to the German High Seas Fleet, which could not risk letting itself be trapped away from its bases by the larger British Battle Fleet.

The high construction and operating costs of Zeppelins limited their number, and the German Navy therefore purchased many more floatplanes for short-range, tactical reconnaissance than it had Zeppelins. As airplane technology advanced and experience was gained using armed, two-seat floatplanes against Allied ships and aircraft, the German Navy realized it needed fighters to defend its naval air stations. The first generation of German naval fighters were single-seat floatplanes and flying

Above: A Brandenburg W29 on patrol in threatening weather. Brandenburgs often patrolled over the North Sea in formations of five aircraft.

boats.

The Navy soon realized that it needed fighters with longer range capable of offensive operations, not just air station defense. A second crewman could help with over-water navigation, use a flexible machine gun for greater combat effectiveness, and provide an additional pair of eyes for scouting for enemies, so a two-seat fighter design was wanted. The challenge was to design a two-seat seaplane fast and maneuverable enough for the fighter role.

The Brandenburg W12 two-seat biplane was a breakthrough design that was just what the Navy wanted. Despite using the same variety of 150–160 hp engines used in the single-seat seaplane fighters, the W12 was as fast and maneuverable as they were, plus it had longer range and endurance and combined these attributes with the greater combat effectiveness of a second crewman with flexible gun. The combat success of the W12 rendered single-seat seaplane fighters obsolete, and many of those on hand were relegated to training or even to storage while the W12 shouldered the burden of combat.

The innovative structural design of the W12 was key to its success. Not only did the W12 sire an extensive family of derivative designs from its parent company, all subsequent German two-seat seaplane fighters from other companies were greatly influenced by the W12, including the all-metal designs of the Junkers and Dornier (Zeppelin-Lindau) companies that were too late for combat.

Only Brandenburg was able to deliver successful new two-seat floatplane fighters in time for combat operations before the armistice. Once the W12 proved its worth over the North Sea, the Navy wanted similar fighters with greater range and endurance, and the larger W19 was developed from the W12 to provide these enhanced capabilities. The Navy also wanted more speed to make intercepting British flying boats and floatplanes easier, and the monoplane W29 derived from the W12 provided greater speed by reducing the drag of the airframe. Finally, the enlarged W33 monoplane provided both the increased speed of the W29 and the longer range of the W19, and all these types were employed effectively on combat operations over the North Sea.

German submarines transited the North Sea en route to and from their patrol areas, and Britain naturally engaged in extensive anti- submarine operations in and over the North Sea to counter the U-boats. These operations included anti-submarine patrols by flying boats, floatplanes, and non-rigid airships, which posed a significant hazard to German submarines. The Germans needed to counter these operations, and the arrival of the Brandenburg fighters made an immediate impact on the situation.

From that time onward British aircraft had to be prepared to defend themselves against effective air attack and the resulting combats were frequent and brutal. Flying from Zeebrugge, Norderney, Borkum, and Ostende on the Flanders coast, the Brandenburgs earned the nickname 'The Hornets of Zeebrugge' for their fierce attacks on Allied floatplanes, flying boats, airships, and surface ships. In addition to downing their share of British aircraft, all six MTBs (Motor Torpedo Boats) involved in an operation on 11 August 1918 were destroyed by the Brandenburgs. Three of the MTBs were sunk by machine-gun fire and the other three MTBs, all in desperate condition, deliberately ran themselves aground in neutral Holland to avoid being sunk, thus saving most of their crewmen.

Naval Aircraft Categories

The German Navy did not use the army aircraft designations; instead the Navy had its own type designation system. Naval aircraft were referred to by individual Marine Number and category. The categories listed at right could be combined to describe the aircraft equipment in greater detail, allowing designations such as C2MGHFT.

Naval Aircraft Procurement

Naval aircraft procurement was controlled by the *RMA* (*Reichs Maritime Amt* – German Admiralty). The prototypes of all new designs were sent to the *SVK* (*See-Flugzeug-Versuchs-Kommando* – Seaplane Experimental and Testing Command) at Warnemünde for testing and evaluation. Seaplanes were built to the standards established in the *Allgemein Baubestimmungen für Seeflugzeuge der*

Naval Aircraft Categories	
Category	**Meaning**
B	Two-seat aircraft with bombing equipment
BFT	Two-seat aircraft with bombing equipment and wireless transmitter
Bu	U-boat aircraft (*U-Boot Flugzeug*)
C2MG	Two-seat aircraft with one flexible and one fixed machine gun
C3MG	Two-seat aircraft with one flexible and two fixed machine guns
CHFT	Two-seat aircraft with one flexible machine gun and wireless transmitter and receiver
CK	Two-seat aircraft with one flexible cannon and one or two fixed machine guns
E	Single-seat flying boat (*Einsitzerboot*)
ED	Single-seat aircraft with twin floats (*Einsitzer mit Doppelschwimmer*)
ED2MG	Single-seat aircraft with twin floats and two machine guns
FL	Aircraft for control of wire-guided attack boats (*Fernienkflugzeug*)
G	Twin-engine aircraft (*Grossflugzeug*)
HFT	Two-seat aircraft with bombing equipment and wireless transmitter
R	Aircraft with three or more engines (*Riesenflugzeug*)
S	Training aircraft (*Schulflugzeug*)
T	Torpedo aircraft (*Torpedoflugzeug*)
U	Practice aircraft (*Uebungsflugzeug für Alleinflieger*)
V	Test aircraft (*Versuchsflugzeug zur Erprobung von Motoren, Propellern, Instrumente, usw.*)

Reichs-marine (General Construction Regulations for Marine Aircraft of the German Navy) and arrived at Warnemünde with current national insignia and assigned Marine Number already applied. Before being accepted for Naval service the aircraft were tested for seaworthiness and flying performance. Problems found during testing had to be resolved before the *SAK* (*See-Flugzeug-Abnahme-Kommission* – Seaplane Acceptance Commission) would approve production.

Fighter Forerunner-FF33L

The main purpose of fighter aircraft is to attack enemy aircraft. Some early fighters like the Vickers Gunbus and SPAD pulpit fighters had a gunner with flexible gun and no gun for the pilot, but the fighter soon evolved and a fixed gun for the pilot became essential. Thus to be included in this work, a fixed gun for the pilot was a requirement, and the C2MG variant of the Friedrichshafen FF33L qualifies. Of the 145 production FF33Ls, 60 were the C2MG category with fixed gun for the pilot; the other 85 were category CHFT with wireless transmitter and receiver and a flexible gun for the observer, but no gun for the pilot. An additional 40 of the C2MG version of the FF33L were ordered but not built and the order was cancelled in December 1918.

The FF33 series was built in greater numbers, 491, than any other German naval aircraft and perhaps any WWI floatplane. The early FF33A and FF33B were unarmed reconnaissance seaplanes built in small numbers. The main early production aircraft was the FF33E, also an unarmed reconnaissance floatplane. The FF33H was an armed development of the FF33E with a flexible gun for the observer and smaller span for better maneuverability. The FF33J was a replacement for the FF33E and the FF33S was a trainer.

The FF33L, the final version of the large FF33 family, was designed for use as an escort and

FF33L Production Orders

Marine Numbers	Category	Qty	Delivery Dates
932–941	C2MG	10	Jan.–Feb./1917
1001–1010	CHFT	10	Mar.–July/1917
1085–1094	CHFT	10	Apr.–June/1917
1117–1126	C2MG	10	Apr.–June/1917
1158–1177	CHFT	20	Apr.–June/1917
1234–1278	CHFT	45	May–Oct./1917
1279–1288	C2MG	10	June/1917
1577–1596	C2MG	20	Aug.–Sep/1917.
3144–3153	C2MG	10	Late 1918
3154–3193	C2MG	40	Cancelled 12/18

Marine #3144 had an experimental tail design.

FF33L Specifications

Engine	150 hp Benz Bz.III
Span	13.3 m
Length	8.825 m
Wing Area	40.54 sq. m.
Empty Wt.	916 kg
Gross Wt.	1,373 kg
Max Speed	136 km/h
Climb to 1,000 m	8 minutes

Right: Friedrichshafen FF33H Marine Number 744 illustrates the aircraft from which the FF33L was derived. The FF33H was already smaller and more streamlined than earlier FF33 models and mounted a flexible gun for the observer. The FF33L featured a spinner and more streamlined nose than the FF33H and a more streamlined tail as well, improving its performance. Addition of a fixed pilot's gun in its C2MG version made the FF33L the transitional design to a real naval fighter.

Right: This rear view of Friedrichshafen FF33H Marine Number 699 shows the older tail surfaces to advantage. These were from the earlier FF33 variants and were replaced with more streamlined surfaces in the FF33L. Friedrichshafen seaplanes were robust, reliable craft that could withstand the rigors of maritime service and had good seakeeping.

patrol fighter and its C2MG variant was the first German seaplane to mount a fixed gun for the pilot. The FF33L was developed from the FF33H and was somewhat smaller and more streamlined than earlier FF33 floatplanes to increase speed and maneuverability compared to its general-purpose ancestors. For a seaplane of the time the FF33L had good maneuverability and handling and fair speed, and it was an effective escort for its unarmed

reconnaissance companions over the North Sea and Baltic.

While its offensive capabilities were modest, the C2MG variant of the FF33L was the transitional design to the floatplane fighter and showed the need for more effective aircraft like the faster, more maneuverable Brandenburg W12 that truly warranted being called fighters through their enhanced offensive combat capability.

Facing Page: Friedrichshafen FF33E Marine #841 *Wölfchen* was the most famous German floatplane. Seen here after its return to Germany and after being restored to a dramatic finish it never wore on operations, *Wölfchen* was the reconniassance carried into the Indian Ocean and back by the German merchant raider *Wolf*, the most successful of all Germany's raiders in WWI. *Wölfchen* played a key role in *Wolf's* success. During the voyage *Wölfchen* was covered in plain, unmarked fabric, and at one point was recovered by captured silk fabric after its original fabric wore out. The FF33E was a reliable maid of all work among German floatplanes and served for most of the war, but only *Wölfchen* became famous.

Above: The FF33L flown by *Oblt.z.S.* Friedrich Christiansen with his personal insignia on the rear fuselage is hanging from the crane at Zeebrugge.

Below: Friedrichshafen FF33L floatplanes lined up on the Mole at Zeebrugge. German seaplanes from Zeebrugge frequently engaged seaplanes and landplanes of the Royal Naval Air Service. Landplanes normally had the advantage in these combats because they were not encumbered with floats, but a number of landplanes were downed by floatplanes.

Above: FF33L C2MG #1123 at left and a companion FF33L, possibly C2MG #1124, at right, are ready for their next mission from Angernsee in 1917. The FF33L was smaller and more streamlined than earlier FF33 variants, giving it better speed and maneuverability. That, coupled with its fixed gun for the pilot, enabled it to undertake more aggressive missions than its antecedents, making it the transition stage to a true floatplane fighter.

Left: Friedrichshafen FF33L Marine Number 1001 was the first production CHFT version and is shown here at the *SVK* at Warnemünde.

Below: The C2MG version of the Friedrichshafen FF33L was the first German naval aircraft with a fixed gun for the pilot and was the forerunner of the seaplane fighter. Here it is shown at the SVK at Warnemünde.

Above: Friedrichshafen FF33L floatplanes from Zeebrugge on patrol over the North Sea sight a Dutch schooner.

Below: One of the Friedrichshafen FF33L floatplanes lands near the Dutch schooner to check it for contraband. German seaplanes captured or sunk a number of vessels like this that they found carrying contraband.

Above: This photo taken at *Seeflugstation Angernsee* in 1917 shows, from left to right, Friedrichshafen FF33L Marine Numbers 1175, 1091, and 1123 in the foreground, and in the right background FF33L Marine Numbers 1236 and 1094. Friedrichshafen FF33E Marine number 805 is in the center background.

Above: Friedrichshafen FF33L Marine Number 1009 is at the far right with FF33L Marine Number 1239 in front of it on the ramp.

Single-Seat Floatplane Fighters

By May 1916 the German Navy decided it needed seaplane fighters to protect its naval air stations and initiated discussions with a number of aircraft manufacturers, including Albatros, Friedrichshafen, Hansa-Brandenburg, Lübeck-Travemünde, Roland, Rumpler, and Sablatnig to build single-seat seaplane fighters. Prototypes were ordered from a number of manufacturers, and the below table lists the aircraft that were built to this requirement. The Brandenburg CC, W17, and W18 were flying boats; the others were floatplanes. Interestingly, all the single-seat floatplane fighters placed in production were derived from production landplane designs.

Single-Seat Seaplane Fighters		
Aircraft	**Number Built**	**Marine Numbers**
Albatros W4	118	747, 785–786, 902–911, 948–967, 1107–1116, 1302–1326, 1484–1503, 1504–1513, 1719–1738
Brandenburg CC	36	946, 1137–1146, 1327–1351
Brandenburg KDW	58	748, 783–784, 912–921, 1067–1076, 1380–1394, 1554–1573
Brandenburg W11	3	988–990
Brandenburg W16	3	1077–1079
Brandenburg W17	1	?
Brandenburg W18	1	2138
Brandenburg W25	1	2258
Friedrichshafen FF43	1	749
L.T.G. SD1	6	1299–1301, 1518–1520
Lübeck-Travemünde F3	1	844 (not accepted by the *SVK*)
Roland W	1	750
Rumpler 6B1	38	751, 787–788, 890–899, 1037–1061
Rumpler 6B2	50	1062–1066, 1188–1207, 1434–1458
Sablatnig SF4	2	900 (biplane), 901 (triplane)
Ursinus	1	782

Specifications of Single-Seat Floatplane Fighters

Spec	KDW	W11	W16	W25	ALB W4 #747	ALB W4 (948–967)	Ru 6B1	Ru 6B2	FF43	SF4 Biplane
Engine	150 hp Benz Bz.III or 160 hp Maybach Mb.III or 160 hp Mercedes D.III	200 hp Benz Bz.IV	160 hp Oberursel U.III	150 hp Benz Bz.III	160 hp Mercedes D.III	160 hp Mercedes D.III	160 hp Mercedes D.III	160 hp Mercedes D.III	160 hp Mercedes D.III	150 hp Benz Bz.III
Span	9.25 m	10.0 m	9.25 m	10.4 m	9.50 m	9.50 m	12.2 m	—	9.92 m	12.0 m
Length	8.0 m	8.2 m	7.35 m	8.8 m	8.50 m	8.50 m	9.05 m	—	8.55 m	8.33 m
Wing Area	28.0 m²	31.4 m²	21.35 m²	36.53 m²	31.0 m²	—	—	—	31.0 m²	28.26 m²
Empty Weight	940 kg.	935 kg.	636 kg.	918 kg.	709 kg	784 kg	854 kg	—	798 kg	798 kg
Flying Weight	1,210 kg	1,215 kg	896 kg	709 kg	989 kg	155 km/h	1,134 kg	—	1,078 kg	1,078 kg
Maximum Speed	170 km/h	176 km/h	170 km/h	160 km/h	155 km/h	160 km/h	152 km/h	166 km/h	163 km/h	158 km/h
Climb to 1,000m	5.9 min.	—	5 min.	6.5 min.	—	5.5 min.	5 min.	—	6 min.	5.5 min.
Climb to 2,000m	14 min.	—	—	—	—	8.5 min.	10 min.	—	12 min.	14 min.
Climb to 3,000m	21 min.	—	27 min.	—	—	23 min.	18 min.	—	—	—
Duration	2.5 hours	—	2 hours	2.5 hours	—	—	—	—	Range: 400 km	—
Guns	1 gun (first 38), 1 or 2 guns (last 20)	2 guns	2 guns	2 guns	1 gun	2 guns	1 gun	2 guns	2 guns	1 gun

Albatros W4

Albatros, the largest WWI German aircraft manufacturer, responded to the Navy's request with the W4, a seaplane development of their Albatros D.I fighter that was being built at the same time. Both the W4 and D.I were powered by the 160 hp Mercedes D.III engine. Due to its floats, the W4 was heavier than the D.I and needed larger wings for more lift. The horizontal tail was larger to compensate for the larger wings, and flight testing of the prototype revealed tail heaviness. Changing the amount of stagger solved the problem and happily also improved climb rate and top speed. The prototype and initial production W4s had ear radiators and a single machine gun. The harsh operating environment caused considerable problems with the wood floats and a number of different designs had to be tried both for strength and seaworthiness. Repairs and water-proofing the leading edge and spars of the lower wings were also required after water damage, including delamination of the spars, was discovered. Starting with the second W4 production batch, two machine guns

Albatros W4 Production Summary

Marine Number	Qty	Notes
747, 785–786	3	Prototypes; 1 gun, ear radiators
902–911	10	1 gun, ear radiators
948–967	20	2 guns, ear radiators
1107–1116	10	2 guns, ear radiators
1302–1326	25	2 guns, ear radiators
1484–1503	20	2 guns, airfoil radiator, 4 ailerons
1504–1513	10	2 guns, airfoil radiator, 4 ailerons
1719–1738	20	As 1504–1513; delivered directly to storage.

Above: *Oblt.zur See* Friedrich Christiansen in the W4 prototype at Zeebrugge. The floats are longer than the initial floats used and have been camouflaged. Christiansen went on to score 13 victories and was awarded the *Pour le Mérite*.

Above: Albatros W4 Marine #964 of the second production batch.

Right: This view of wrecked Albatros W4 Marine #958 of the second production batch being recovered from the water shows the three-color naval hexagonal camouflage fabric used on later production aircraft. The printed fabric was used on all upper surfaces. Printing the color on the fabric during manufacture saved time, labor, the weight of paint, and the paint itself.

were fitted.

The arrival of summer weather revealed problems with the ear radiators, which had to be replaced with a new design. Starting with the fifth production batch (1484–1503), airfoil radiators were fitted to reduce drag. To improve maneuverability the last two production batches were fitted with ailerons on all four wings.

By the time the final production batch was built, the Brandenburg W12 two-seat floatplane fighter had proved it was more effective in combat than single-seater floatplane fighters and the final production batch of W4s was delivered directly to storage. Eight W4s were traded to the Austro-Hungarian Navy in July 1918 in return for Austro-Daimler V-12 engines to power Staaken R-planes. Designated E5 to E12 in Austro-Hungarian service, these saw no combat. By August 1918 only four W4s were on combat duty on the North Sea and another five were on combat duty in Turkey.

A total of 118 W4 fighters were built in a series of production batches summarized in the table above.

The prototype Albatros W4 Marine #747 as built with stained wood fuselage and clear-doped linen flying surfaces. After it was assigned to Zeebrugge it was partly over-painted in camouflage colors.

Left: Albatros W4 #956 of the second production batch. The ear radiator is prominent and, located on both sides of the aircraft, partially spoiled the clean aerodynamic lines of the fuselage. This was addressed by using lower-drag airfoil radiators in the last three production batches.

Above: Albatros W4 Marine #965, an aircraft of the second production batch.

Below: This photo shows a woman in the cockpit of an Albatros W4; the photo seems to have been retouched to include her in 'pin-up' style, perhaps for a postcard. The W4 appears to be Marine #965 shown above; the dent in the spinner appears to be the same.

Albatros W4 Marine #911 after tactical markings were added. Close inspection of available photos shows the circular markings were in three colors, thought to be the German national colors, and were applied to the top of the rear fuselage in addition to the sides. This fighter was flown by *Lt.s.Z.* Schulz, who downed three aircraft while flying it.

Right: A pair of Albatros W4 fighters ready for launching; Marine #911, last aircraft of the first production batch, is nearest the camera. W4 #911 later had additional tactical markings added as shown in the photo below.

Below: Seaplanes lined up at Windau in 1917 include at least seven Albatros W4 fighters, with Marine Numbers 911, 958, and 1107 visible on the original photo. Albatros W4 Marine #911 is the aircraft in the right background with the two circles on the fuselage. Marine #911 was the most successful Albatros W4 in the Baltic; *Lt.z.S.* Schulz downed three aircraft while flying it. Friedrichshafen FF33L Marine #1263, a category CHFT reconnaissance two-seater, is in the left foreground, and a similar machine is at right.

Albatros W4 Marine #1512 was the next to last W4 built in the next to last production batch. It represented the final W4 production configuration with two guns, an airfoil radiator, and ailerons on all four wings for better maneuverability, and was finished in standard naval late-war camouflage.

Above: Albatros W4 Marine #1318, from the fourth production batch, is shown here at the Austro-Hungarian *Seeflugstation Puntisella* in July 1918 before its new Austro-Hungarian serial E12 was applied.

Left: Albatros W4 Marine #1322 from the fourth production batch has been damaged by a hard landing. It has the late insignia.

Above: W4 Marine #1486 having a bad day.

Above: Albatros W4 Marine #965. This photo appears to be the one used to create the 'pin-up' photo on page 17.

Above: W4 #1486 of the fifth production batch before its accident shows the final production configuration of four ailerons, an airfoil radiator, and the late naval camouflage scheme.

Left: W4 #1512, sixth batch, also shows the final production configuration of four ailerons, an airfoil radiator, and the late naval camouflage scheme.

Brandenburg KDW

Above: This striking view of Marine #748, the first prototype, shows the washed out ailerons. Many WWI airplanes had the angle of attack of their ailerons reduced or 'washed out' to ensure the inner part of the wing stalled before the outer part, where the ailerons were fitted. With no 'wash out', the outer part of the wing would stall first and the pilot would lose aileron control at the beginning of the stall, making a spin much more likely. The metal fairings enclosing the apex of the wing star-struts of #748 are not seen on later aircraft.

The Brandenburg KDW (*Kampf Doppeldecker Wasser* – literally 'combat biplane water') was developed from the earlier Brandenburg KD (*Kampf Doppeldecker*) landplane fighter. The Brandenburg KD was designed in Germany in 1916 and went into production for the *Luftfahrtruppe*, the Austro-Hungarian air service, as the Brandenburg D.I fighter.

The Brandenburg KD, or D.I, was basically a conventional, wire-braced wooden design of the time except for one notably different feature, the interplane struts. The metal interplane struts, streamlined by laminated wood fairings, were designed as a set of four triangles that met at their apex, giving a unique star appearance. The star-strut design was very strong and eliminated the need for the drag-producing bracing wires featured by contemporary designs. The resulting Brandenburg KD was a strong, fast fighter. Unfortunately, its single machine gun was mounted in a streamlined housing above the wing, making it impossible for the pilot to clear jams during flight. Worse, the KD lacked both maneuverability and stability, and was prone to stalling and spinning with little provocation.

The German Navy's need for a seaplane station defense fighter led to development of the Brandenburg KDW, a straight-forward conversion of the KD to a floatplane fighter. Floats replaced the wheels and both the wing span and area were increased to provide additional lift to compensate for the extra weight of the floats. Fortunately, a synchronized machine gun was mounted on the starboard side of the fuselage instead of the clumsy, over-wing mounting used by the KD. Initially the gun was mounted too far forward to be reached by the pilot in case of a jam during flight, but this was rectified with the first production series. The final production series was intended to mount two guns adjacent to the cockpit where the pilot could reach them in flight, but some machines carried only one gun.

The three prototype aircraft had the 150 hp Benz Bz.III engine with car-type, frontal radiator. Aircraft of the second production series had the 150 hp Benz Bz.III (if intended for the Baltic) or the 160 hp Mercedes D.III engine with radiator in the starboard side of the upper wing (if intended for Flanders or the North Sea). All aircraft of the first, third, and fourth

Above: The KDW prototype before armament was fitted. No Marine Number is visible and the frontal, car-type radiator and overall aerodynamic cleanliness of the design are prominent.

Left: KDW #748, the machine used by *Lt.d.RMl* Fritz Hammer at Angernsee to bring down Sikorski *IM-6* on 23 Sept 1916. The basic sturdiness of the design was tested during Hammer's running combat, in which he made four separate passes at the Russian bomber, once nearly losing control when he was caught in the turbulence of the bomber's wake. Note the position of the machine gun, which was so far forward that the muzzle can be seen close to the radiator.

Left: Hammer criticized the inaccessibility of the gun on the starboard side of #748, clearly out of reach, after his combat with the Sikorski. According to Hammer he might have been able to clear the simple jam that occurred on his fourth pass had he been able to reach this gun.

Right: The prototype KDW Marine #748, as confimed by the fairings over the apex of the struts. The Marine Number has not yet been applied.

production series had the 160 hp Maybach Mb.III engine with radiator in the starboard side of the upper wing. In the summer of 1917 early production aircraft were refitted with additional small interplane struts bracing the upper wing to increase torsional stiffness of the long-span (compared to the KD) wing, thereby improving aileron response. From the third production batch these struts were installed during production.

Based so closely on the KD, the KDW inherited its performance and flying qualities. These included structural strength and good speed for a floatplane fighter, although the additional weight and drag of the floats and longer wings necessarily reduced speed and climb given that engine power was essentially the same. Less happily, the KDW also inherited the poor stability and flying qualities of the KD. A number of changes to the vertical tail surfaces were made during production in an attempt to improve these, but without notable success. As a result, the KDW was able to engage reconnaissance seaplanes and similar targets with a reasonable chance of success, but was at a distinct disadvantage in combat with contemporary landplane fighters. The KDW's most significant operational success was when the first prototype, Marine Number 748, downed Russian Ilya Mouromets *IM-6*, a four-engine reconnaissance-bomber, one of only three downed in air-to-air combat during the war.

The KDW was delivered in small batches from September 1916 to February 1918. Only about 2,365 aircraft were produced for the German navy during the war, so the 58 KDW aircraft produced represented a reasonable success. However, the competing Albatros W4, the floatplane conversion of the successful Albatros D.I fighter, was produced in twice the numbers of the KDW, 118 being delivered from September 1916 to December 1917. Although the KDW with 160 hp Mercedes D.III had a slight speed advantage over the Albatros W.4, the KDW

KDW Production		
Marine Number	**Qty**	**Notes**
748, 783, 784	3	Prototypes. 1 gun, 150 hp Benz Bz.III engine. Completed September 1916.
912–921	10	1 gun, 160 hp Maybach Mb.III engine. Series completed February 1917.
1067–1076	10	1 gun, either 160 hp Mercedes D.III engine (aircraft destined for the Western Front) or 150 hp Benz Bz.III engine (aircraft destined for the Baltic). Series completed in March and April 1917.
1380–1394	15	1 gun, 160 hp Maybach Mb.III engine. Series completed in summer 1917.
1554–1573	20	1 or 2 guns, 160 hp Maybach Mb.III engine. Series delivered between October 1917 and February 1918.

with the 150 hp Benz was slower. Furthermore, the W.4 had better climb rate and maneuverability, better visibility from the cockpit, and, perhaps most important, much better flying characteristics. Both types served on the Flanders front, the North Sea, and in the Baltic. Eventually the KDW succumbed to the same poor flying qualities and cockpit vision problems as its landplane predecessor, the D.I.

Above: Marine Number 783, the second KDW prototype, showing its frontal radiator. The machine gun is far forward, out of the pilot's reach in case of jams. The wing structure is evident through the fabric. The *Balkenkreuz* on the wings appear to be painted over white backgrounds. Although it is difficult to see in this view, there is no fixed fin above the fuselage.

Above: Marine Number 783, the second KDW prototype, clearly shows the forward mounting of the machine gun in this photo. Although the view of the rudder is washed out, the only fixed fin is shown below the fuselage.

Above: The inaccessible location of the machine gun is also clearly evident on the second KDW prototype, Marine #783. This aircraft was being ferried from Warnemünde to Windau by *Lt.z.S.* Joachim Coeler when it was forced down near Memel on 23 Sept., 1916. This aircraft may not have completed its journey, for it does not appear in subsequent flying entries at Windau. In March, 1917, Coeler was ordered to the Putzig *Seekampfeinsitzerschule* to be its commander. KDW Marine #748 is in the right background, identified by the metal fairings at the apex of its interplane struts.

Above: Side view of Marine Number 912, the first production KDW, shows the machine gun has been moved back to enable the pilot to reach it in flight to clear jams. This change was made on the basis of early combat reports on #748.

Above: The first KDW from the initial production batch, Marine #912 went to Zeebrugge. On 10 May, 1917, this aircraft was wrecked while being flown by *Oblt.z.S* Kurt Reinert, who later died from the injuries he sustained. Note that that *Lt.* Hammer's remarks were heeded; the machine gun is now located closer to the cockpit. Also the radiator has been moved to the upper wing, just right of center.

Above: Another KDW from the first production batch was Marine #921, which initially went to Libau, arriving there on 24 Feb., 1917. This aircraft was apparently at Putzig by May, and was wrecked while being flown by *Lt.* Markwald on 11 August, 1917. This photo was taken in front of a Hansa-Brandenburg hangar, apparently prior to June, 1917, when additional bracing struts were ordered to be installed on all KDWs.

Above: A late production KDW, Marine Number unknown, displays the fixed upper fin added to later aircraft to improve stability and the additional small interplane struts added for better aileron effectiveness. Compare this aircraft with #1562 of the last production batch; this aircraft retains the plain finish of early production KDWs and the intermediate location of the guns, so may be of the third production batch.

Below: This appears to be another photo of the KDW above, and provides an interesting view of the star struts augmented with the auxiliary bracing struts. This arrangement was strong and eliminated the need for drag-producing bracing wires, but was heavier than conventional wing bracing. The additional weight and drag of the auxiliary bracing struts basically eliminated the advantages of the star-strut design, and later designs returned to conventional wing struts.

Above & below: These photos, Marine Number unknown, clearly shows the light interplane struts ordered to be added to all KDWs in June, 1917. The light struts run from the front spar of the lower wing to both spars of the upper wing and increased the torsional stiffness of the upper wing, which improved aileron response. Compared to the prototypes, the machine gun has been moved back closer to give the pilot better access in case of jams. A fixed vertical fin has been added above the fuselage, which was not done on early production batches.

Above: Marine Number 1562 of the final KDW production batch displays the additional interplane struts and shows off the three-color hexagonal camouflage fabric used on the upper surfaces of German naval aircraft late in the war. The machine guns have been moved up in front of the pilot for better access.

Above: A KDW from the last production batch, #1562 was one of the aircraft originally sent to *II.Seefliegerabteilung*, presumably for assignment to a western front unit. Following a series of crashes at the *Seekampfeinsitzerschule* at Putzig, this and several other *II.SFA* machines were ordered to be transferred to *I.SFA* and sent to the Putzig school. Standard late naval lozenge camouflage was applied to this machine. The fixed vertical fin above the fuselage is enlarged compared to the prototypes and early production machines.

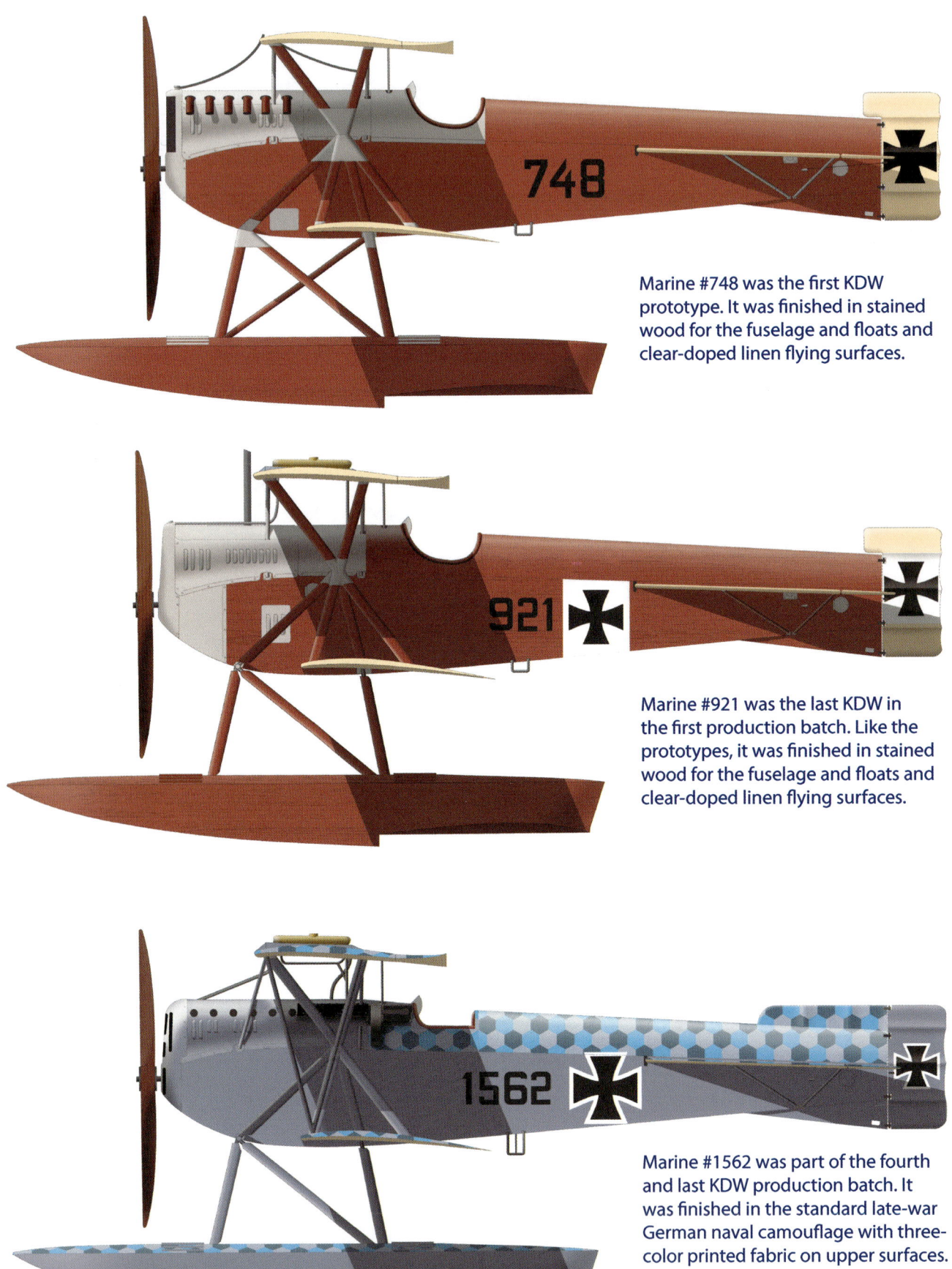

Marine #748 was the first KDW prototype. It was finished in stained wood for the fuselage and floats and clear-doped linen flying surfaces.

Marine #921 was the last KDW in the first production batch. Like the prototypes, it was finished in stained wood for the fuselage and floats and clear-doped linen flying surfaces.

Marine #1562 was part of the fourth and last KDW production batch. It was finished in the standard late-war German naval camouflage with three-color printed fabric on upper surfaces.

Brandenburg W11

Above: W11 Marine Number unknown; this type was a slightly enlarged KDW with two machine guns and the more powerful 200 hp Benz Bz.IV engine. Only three aircraft, marine numbers 988–990, were built in late 1916 due to the marginal improvement over the KDW. Of the three aircraft built, at least two, Marine Numbers 988 and 989, were assigned to *Flandern 1* at Zeebrugge.

Following the KDW, Brandenburg constructed three other single-seat floatplane designs during the war. Like the final KDW production series, all were armed with two fixed, synchronized Spandau machine guns.

The first design developed from the KDW was the Brandenburg W.11, an enlarged KDW powered by a 200 hp Benz Bz.IV engine. Span was enlarged slightly to 10.0 m, length to 8.2 m, and wing area to 31.4 sq. m. The additional power increased speed to 176 km/h (109 mph), but flight characteristics were not improved. With the increased power an increased climb rate would also be expected, but no data survive to confirm that. Only three aircraft, marine numbers 988–990, were built in late 1916 due to the marginal performance improvement over the KDW and the great success of the Brandenburg W12 two-seat fighter.

Right: The logo of the German Anker propeller company on an Anker propeller.

Brandenburg W16

The W16 of 1916 was distinctly different from the KDW and was powered by an Oberursel U.III rotary engine of 160 hp. Span was the same as the KDW and due to the rotary engine it was much lighter. Flying characteristics likely were somewhat improved over the KDW, but speed was the same and above 1,000m climb was inferior. Like the W11, only three were built, Marine Numbers 1077–1079, because they were no improvement over the KDW, and there is no record the W16 flew on operations.

This Page & Facing Page:
The W16 was an attempt to create a replacement for the KDW that had better maneuverability and handling qualities. It had a larger fixed fin to improve stability and the heavy star-strut arrangement was replaced with a new design that did not obstruct the pilot's view as much. Despite the same power and its much lighter weight than the KDW, it was no faster and its climb rate was actually lower above 1000m. The massive spinner was an attempt to minimize drag. Only three W16 fighters were built.

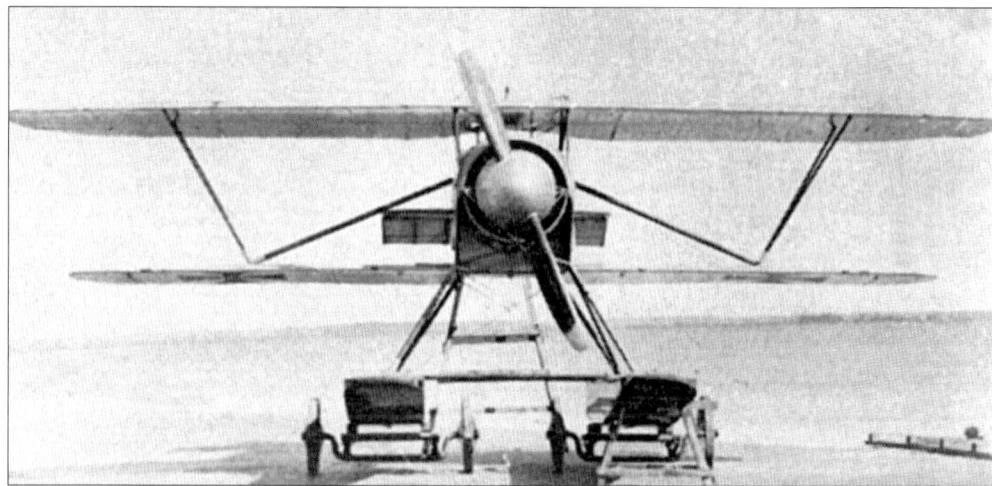

Brandenburg W25

Above & Below: The final development of the KDW configuration was the W25. It has three-color naval hexagonal fabric on the upper surfaces; conventional interplane struts replaced the 'star-strut' arrangement of the KDW.

The final Brandenburg single-seat floatplane fighter design was the W25, a development of the KDW with conventional wing bracing. The W25, with its revised wing bracing, ailerons on all four wings, and increased wing span and area, was clearly an attempt to improve on the maneuverability, flight characteristics, and field of view of the KDW. Weight was reduced slightly due to replacement of the star struts, but with the same 150 hp Benz Bz.III engine and the larger wing with its greater drag, both speed and climb rate were actually less than the KDW. Due to its inferior performance compared to the KDW only a single aircraft, Marine Number 2258, was produced.

This Page: The W25 had the fixed upper vertical fin of the late-production KDW, but differed in having ailerons on all four wings, which also featured increased span and area. These changes likely improved its handling characteristics, but not enough for a production order.

Rumpler 6B1 & 6B2

Above: Rumpler 6B1 Marine #1045 of the last 6B1 production batch shows its propeller spinner and streamlined nose in this air-to-air photograph. Production 6B1 aircraft had the Rumpler C.IV's curved horizontal tailplane. Surprisingly, the 6B1 was more effective on operations than either the Albatros W4 or KDW despite being derived from a two-seat reconnaissance airplane. Pilots liked its excellent handling qualities and its performance was competitive.

The Rumpler 6B1 (known to the Navy as the Rumpler ED) was derived from the successful Rumpler C.I two-seat reconnaissance aircraft. The weight of the floats was compensated for by elimination of the observer and his equipment, and the upper wing was moved forward to compensate for the forward shift in center of gravity. Like the Rumpler C.I and Albatros W4, the 6B1 was powered by a 160 hp Mercedes D.III. The 6B1 retained the C.I's single fixed gun for the pilot. The prototype, Marine #751, was delivered to the seaplane testing command on 7 July 1916. After testing it was accepted on 10 August, and the first production batch of ten fighters was ordered on 14 August, followed by a further batch of 25 fighters.

Rumpler 6B2

The Rumpler 6B1 was followed in production by the improved 6B2 that had the ability to mount two machine guns. The 6B2 also had refined aerodynamics derived from later production Rumpler C.IV aircraft that gave it greater speed; specifically the spinner was replaced by a rounded nose. Three production batches totaling 50 aircraft were ordered in early 1917. However, production was slow; the first aircraft were delivered a full ten months after being ordered, and only about half were fitted with two guns, the others mounting a single gun. Production data and other information about use suggest that by the time the 6B2 was ready, the Navy strongly preferred two-seat seaplane fighters, reducing the urgency for the 6B2 and other single-seat fighters. The 6B2 therefore saw limited operational service and was mostly used for training. Four 6B2 aircraft were sold to the Austro-Hungarian Navy, which assigned them serial numbers E1–E4.

Above: The first prototype Rumpler 6B1, Marine#751, is photographed here conducting flight trials at the Rumpler facility on the Müggelsee near Berlin in June 1916. The armament has not yet been installed.

Above: The prototype Rumpler 6B1 while at the *SVK* in Warnemünde. The light color does not contrast well with the sky background, which was good for air-to-air camouflage and bad for photography. Power was provided by a 160 hp Mercedes D.III engine. The first three prototypes had the C.I horizontal stabilizers with straight leading edges; production aircraft had the C.IV tailplane with curved leading edges.

Rumpler 6B1 & 6B2 Production

Marine Numbers	Qty	Notes
751, 787–788	3	Prototypes, C.I tailplane
890–899	10	First 6B1 production, C.IV tailplane
1037–1061	25	Second 6B1 production
1062–1066	5	6B2 pre-production
1188–1207	20	First 6B2 production
1434–1458	25	Second 6B2 production

Above & Below: Rumpler 6B1 Marine #1045 photographed in flight. This fighter was stationed at Borkum on September 13, 1917, where these photos and the one starting this section were probably taken. The photo below shows the camouflage on the upper surfaces.

Left: Rumpler 6B1 Marine #1045 displays its agility for the photographer.

Above: The prototype Rumpler 6B1 photographed at the *SVK* at Warnemünde, where the aircraft was delivered on July 7, 1916. It was accepted on August 10, 1916, and this was quickly followed by a production order for the first batch of 10 aircraft on August 14, before the next two prototypes, 787 and 788, were completed. Someone has drawn in the outline of the fin and rudder on the original print due to the low contrast between the sky and the aircraft.

Below: Rumpler 6B1 W4 Marine #751 is shown after being repainted in camouflage colors after assignment to Zeebrugge. The straight leading edge of the C.I-style tailplane is clearly shown. Flying from Zeebrugge, *Lt.z.S.* Neimeyer used Rumpler 6B1 #751 to down a Short Seaplane on 31 Aug. 1916 and a Caudron G.4 on 7 Sept. 1916.

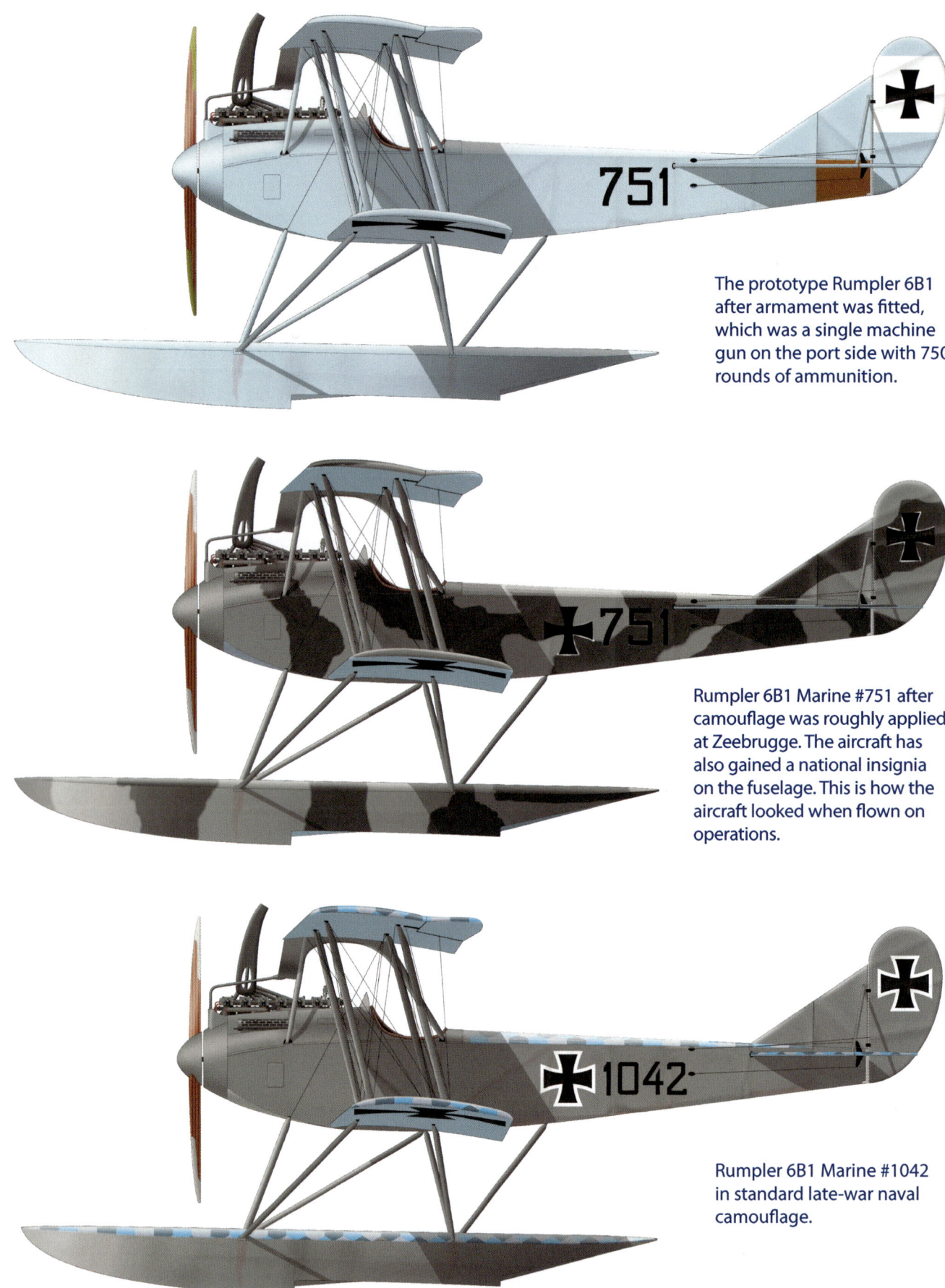

The prototype Rumpler 6B1 after armament was fitted, which was a single machine gun on the port side with 750 rounds of ammunition.

Rumpler 6B1 Marine #751 after camouflage was roughly applied at Zeebrugge. The aircraft has also gained a national insignia on the fuselage. This is how the aircraft looked when flown on operations.

Rumpler 6B1 Marine #1042 in standard late-war naval camouflage.

Above: Rumpler 6B1 Marine #899 was the last aircraft of the first production batch. It is seen here being hoisted by a crane for launching, something every German seaplane was equipped to do.

Below: Rumpler 6B1 Marine #899 after successful launch. The curved tailplane applied to production aircraft is clearly visible. The standard late-war, three-color lozenge fabric does not appear to have been used on this aircraft; instead, the upper surfaces of the wings appear to be sprayed in two colors. On 15 May 1917 #899 was stationed at Borkum on the Flanders coast for operations over the North Sea.

Above: Rumpler 6B1 W4 Marine #1051 shows its dark coloring and operational markings. The 6B1 and 6B2 had good performance despite their size and had far better handling qualities than the KDW.

Above: Rumpler 6B1 Marine #1042 of the second production batch is launched into the water and is seen just leaving the launching dolly. This aircraft was assigned to Konstanza on May 15, 1017, suggesting this photograph was taken there.

Rumpler 6B1 Marine #1051 of the
second 6B1 production batch;
see photo on facing page.

Above: Rumpler 6B1 Marine #1059 serves as a backdrop for a group photograph; unfortunately, the identities of the men and dogs are not known. Either this is a very informal photograph or the photographer made this shot while setting up for the formal portrait. The Marine Number has been painted on the side of both floats to identify the aircraft, and the Axial propeller has a metal tip to reduce erosion from spray. Aviators seem to like dogs.

Above: Rumpler 6B2 Marine #1066 was from the 6B2 pre-production batch. From the rear it is difficult to tell the 6B1 from the 6B2. Although the 6B2 was designed to carry two machine guns, only about half the aircraft actually had two guns.

Above: Rumpler 6B2 floatplanes 1439, 1062, 1063, and 1194 are lined up at either Kiel Holtenau or further east at one of the Baltic *Seeflugstationen* (naval air stations). Marine #1439 carries a large white circle as a tactical identification marking, while #1062 carries two white circles and #1063 carries three white circles.

Right: This Rumpler 6B1 has been repainted in the late, straight- sided national insignia specified by the Navy on March 30, 1918. It is being readied for a training flight at Putzig naval air station.

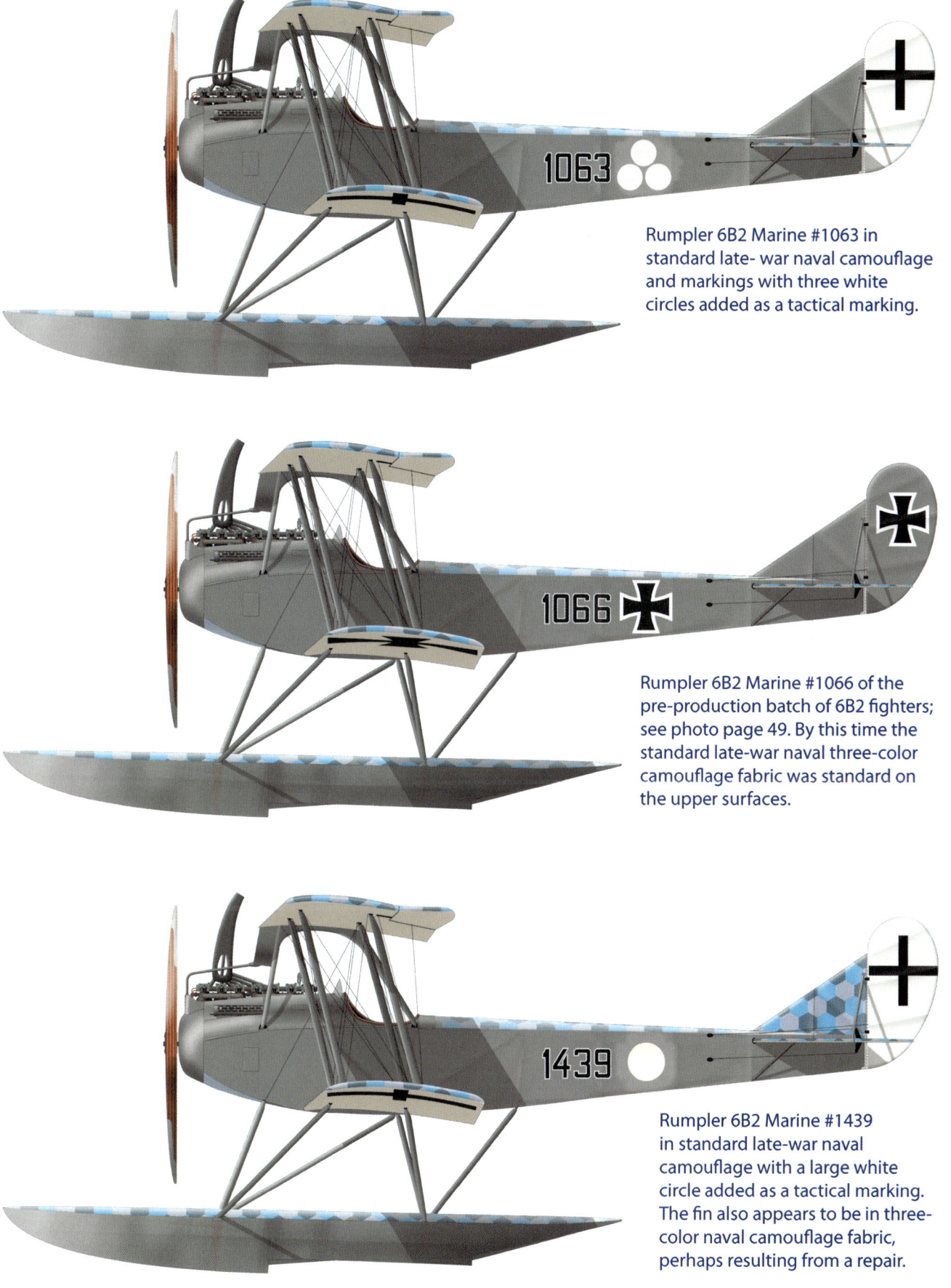

Rumpler 6B2 Marine #1063 in standard late- war naval camouflage and markings with three white circles added as a tactical marking.

Rumpler 6B2 Marine #1066 of the pre-production batch of 6B2 fighters; see photo page 49. By this time the standard late-war naval three-color camouflage fabric was standard on the upper surfaces.

Rumpler 6B2 Marine #1439 in standard late-war naval camouflage with a large white circle added as a tactical marking. The fin also appears to be in three-color naval camouflage fabric, perhaps resulting from a repair.

Above: Rumpler 6B2 Marine #1189 was from the first 6B2 production batch. Its lack of propeller spinner was due to research to improve the Rumpler C.IV series that showed a streamlined nose without spinner had less drag. This made the 6B2 14 km/h faster than the similarly-powered 6B1, a significant improvement for such an apparently minor change. Here is is shown on its beaching gear at *Seeflugstation Puntisella* on July 9, 1918, in Austro-Hungarian service.

Left: Rumpler 6B2 in Dutch service post-war.

Right: An Austro-Hungarian Rumpler 6B2 flies ove rthe harbor at Trieste.

Above & Below: Austro-Hungarian Rumpler 6B2 E3 (formerly Marine #1090) is moored next to the officers' mess at Igalo naval air station. The upper surfaces of the wings are in the German naval three-color printed camouflage fabric, but unusually there are no national insignia on the upper wings.

Friedrichshafen FF43

Above: The prototype Friedrichshafen FF43 on the ramp of the airship hall at the Friedrichshafen works. The assigned Marine Number, #749, has not yet been applied. **Below:** FF43 front view shows the FF43 was well streamlined for its time.

Friedrichshafen was the largest supplier of seaplanes to the German Navy, and its products were noted for being robust and having good seakeeping qualities. The Friedrichshafen FF43, Marine #749, was an original seaplane fighter design powered by a 160 hp Mercedes D.III that first flew on 30 September 1916 but remained a single prototype.

Mounting two machine guns, it had a maximum speed of 163 km/h. This was slightly faster than the contemporary Albatros W4 that was selected for production, so the FF43 must have had other shortcomings, and one source notes that it had a poor rate of climb, which is probably why it was not selected for a production order.

Above: The FF43 under evaluation at the *Seeflug-Erprobung* (Seaplane Test Facility) at Warnemünde. With better speed than the Albatros W4 and two guns to the W4's single gun (until later production W4 batches), the FF43 seemed to offer as much potential as the W4. However, the FF43 was not chosen for production, perhaps due to a poor rate of climb.

Above: The FF43 prototype with Marine Number and national insignia applied. Despite remaining a single prototype, the FF43 was assigned to Zeebrugge for operational service. On November 10, 1916, *Lt.z.S.* Schuler shot down one of two Short seaplanes he encountered during a coastal patrol between Zeebrugge and Dunkirk while flying the FF43.

Above: The FF43 was a well-streamlined aircraft of good proportions that looked a likely prospect for production.

Above: The FF43 without engine cowling shows engine details and the right gun mount.

Left: The FF43 with wings removed shows details of the float struts.

Above: FF43 outside the airship hall before markings were applied.

Right: The FF43 with wings removed shows details of the float struts.

Below & Right: The FF43 at the airship hall before its Marine Number was applied.

Roland W

Above: The Roland WD photographed at the L.F.G. factory.

The Roland W, Marine #750, was derived from the operational Roland C.II two-seater via the D.I fighter. Like the C.II and D.I, the W was powered by a 160 hp Mercedes D.III and mounted a single fixed machine gun, but other details are not known and no production was undertaken.

The Roland C.II and D.I were more noted for their speed than good handling qualities, and it is likely the similar Roland W did not excel in handling either, with the result the production order went to the more docile and maneuverable Rumpler 6B1.

Above: The Roland W, Marine #750, was derived from the Roland C.II two-seat reconnaissance airplane.

Above: Except for its ear radiators, the Roland W was nicely streamlined. It used the *Wickelrumpf* technology of fuselage construction, narrow ribbons of wood wrapped around a mold and glued into a strong, light-weight shell that was streamlined and resistant to combat damage. *Wickelfumpf* was pioneered by L.F.G. and later used successfully by Pfalz.

Below: The Roland W photographed at Warnemünde.

Sablatnig SF4 Biplane & Triplane

Above: The sole Sablatnig SF4 biplane prototype was Marine Number 900.

Designed as a single-seat floatplane fighter, the SF4 was unique in that it was built in both biplane and triplane versions. Both were powered by a 150 hp Benz Bz.III and carried one fixed, forward-firing Spandau machine gun. The biplane, Marine #900, was tested first, and while speed was competitive, it had the lowest climb rate of all the single-seat floatplane fighter competitors. Worse, its maneuverability was poor due to its large wingspan and, despite its multitude of bracing wires, its wing structure was insufficiently robust; wing vibration was excessive in even a shallow dive. To improve the climb rate a triplane version, Marine #901, was built; like the biplane it was not competitive.

Company founder Josef Sablatnig was a trained mechanical engineer and a well-known pioneer pilot, so it is especially disappointing that he was unable to design an airframe that was at once light, strong, and streamlined. Although nose entry was streamlined, the wing structure created a lot of drag due to extensive bracing wires, and despite that the wing structure was weak.

Above & Facing Page, Bottom: The SF4 biplane prototype is photographed in the snow; the broad interplane struts may have helped streamlining but obstructed the pilot's field of view to the sides.

Below: The SF4 triplane carried Marine #901 and had ailerons on all wings. Little information is available on the triplane SF4, but it was not selected for production.

Ursinus U.1

Above: The Ursinus U.1 floatplane fighter was unique in having retractable floats, a significant innovation for its time. To improve maneuverability the 150 hp Benz Bz.III engine was located on the center of gravity.

The most innovative single-seat floatplane fighter prototype built was the Ursinus U.1, which had retractable floats for higher speed through reduced drag. To improve maneuverability the engine was located on the center of gravity and drove the propeller via an extension shaft. The floats were retracted by a manual crank. During trials there were problems with the propeller extension shaft and float retraction mechanism. The prototype was destroyed before it could achieve its estimated top speed of 200 km/h. Its Marine Number was 782.

Above: Front view of the Ursinus floatplane fighter shows its clean lines even with the floats extended. The aircraft was built by Gotha to Ursinus's design, and was known within Gotha as the Gotha WD10.

Above: The inside upper edges of the retractable floats were beveled to fit closely against the fuselage, which was shaped to accommodate them when retracted.

Above: The Ursinus floatplane fighter Marine Number 782 seen from the side with floats extended. The cockpit was well aft because the engine was set back to the center of gravity and drove the propeller via an extension shaft.

Below: The Ursinus U.1 was an innovative design attempt to reduce the drag penalty of floats.

L.T.G. FD1

Left: Three FD1 prototypes, Marine #1299–1301, were ordered. Marine #1299 shown here was destroyed during static load testing. Problems noted during testing resulted in an order for three redesigned versions, at least two of which were delivered to the SVK, but no production resulted.

The Luft Torpedo Gesellschaft, Johannisthal, normally designed aerial torpedoes. Nevertheless, L.T.G. built Marine #1299 and delivered it to Warnemünde, were it was destroyed during load testing. Three re-designed aircraft were ordered, and the first two were delivered to the *SVK* in July 1918. The type was still being tested when the war ended. The engine was listed as a 150 hp Benz Bz.III. However, the hand (direction of rotation) of the propeller is opposite to normal German engines, including the Benz, indicating the engine must have been geared, but details are not known.

Below: FD1 Marine Number 1518 was the first of three of the re-designed, strengthened FD1 fighters ordered; the vertical tail was greatly enlarged compared to the first version. It was delivered to the *SVK* in July 1918 together with Marine #1519. Testing was incomplete when the war ended. It is difficult to understand why an experimental single-seat floatplane fighter was being tested this late in the war after the great success of the Brandenburg two-seat floatplane fighters. Perhaps this aircraft was tested more to evaluate its technology than as a potential production fighter.

Flying Boat Fighters

Above: Marine #946, the Brandenburg CC prototype, was armed with one gun and had a frontal radiator.

In addition to the single-seat floatplane fighters discussed above, Brandenburg also designed a series of single-seat flying boat fighters. These were an anomaly because the German Navy generally preferred floatplanes to flying boats.

The Brandenburg CC was the only flying boat fighter to serve on operations with the German Navy. Flying boats were not popular with German pilots flying from the cold, rough waters of Northern Europe and the CC was soon withdrawn from German service. However, the Brandenburg CC and its more powerful development, the W18,

were popular with the Austro-Hungarian Navy, which enjoyed warmer waters than Germany. The Austro-Hungarian Brandenburg CC fighters had more powerful engines than the German original and were notably faster; they served successfully in the Adriatic as did their replacement, the W18. However, in 1918 the Austro-Hungarian Navy started procuring faster Phönix D.I land-based fighters to supplement the W18s, and the Phönix fighters took an ever-greater role in defending naval air stations in the Adriatic.

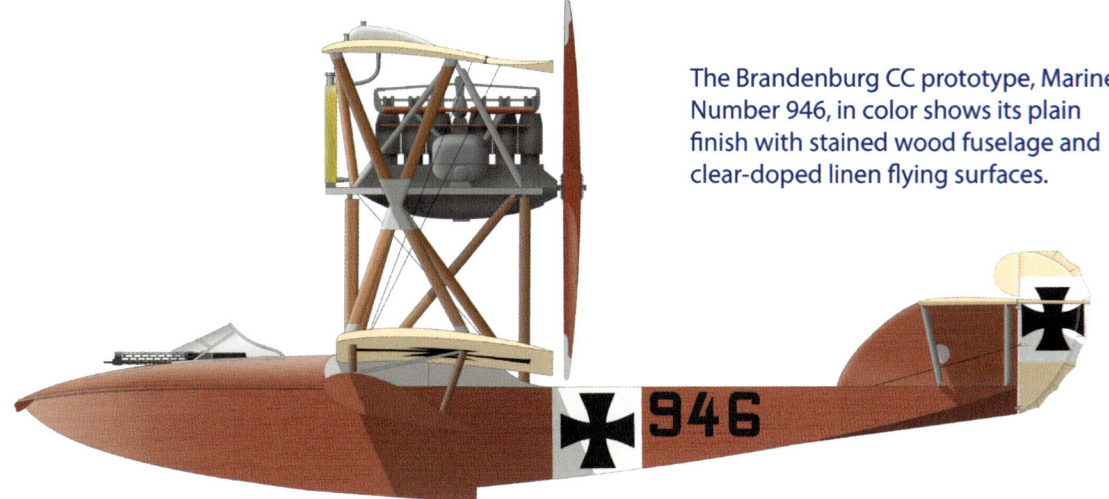

The Brandenburg CC prototype, Marine Number 946, in color shows its plain finish with stained wood fuselage and clear-doped linen flying surfaces.

Specifications For Flying Boat Fighters				
Type	Brandenburg CC (German)	Brandenburg CC (Austro-Hungarian)	Brandenburg CC (Austro-Hungarian)	Brandenburg W18 (Austro-Hungarian)
Engine	150 hp Benz Bz.III	185 hp Austro-Daimler	200 hp Hiero	230 hp Hiero
Span	9.50 m	9.3 m	9.3 m	10.7 m
Length	8.50 m	7.65 m	7.65 m	8.64 m
Wt. Empty	709 kg	716 kg	800 kg	812 kg
Wt. Loaded	989 kg	1,030 kg	1,030 kg	1,092 kg
Max. Speed	155 km/h	170 km/h	180 km/h	180 km/h
Climb to 1,000 m	5.5 min.	5 min.	4 min.	5 min.
Climb to 2,000 m	8.5 min.	11.2 min.	—	11.2 min.
Climb to 3,000 m	23 min.	—	16 min.	23.4 min.
Armament	1–2 guns	1 gun	1–2 guns	2 guns

Brandenburg CC with two guns and airfoil radiator, the standard configuration.

Brandenburg CC

Above: The compact, streamlined Brandenburg CC was the only flying boat fighter used by the German Navy.

The Brandenburg CC shared the star-strut wing design of the Brandenburg KDW. The star-strut design was stronger but slightly heavier than conventional bracing and eliminated the need for external bracing wires. The prototype and German production CC aircraft were powered by a 150 hp Benz Bz.III. The prototype had a single fixed gun for the pilot; production aircraft mounted two fixed guns and a frontal radiator. Like later KDW fighters that shared the star-strut bracing, additional interplane struts to stiffen the wingtips for improved aileron response were added to some CC fighters.

Although it was placed in production, with 35 built in addition to the prototype, German pilots did not consider the Type CC suitable for flying in Northern Europe, perhaps because as a flying boat it did not keep the pilot as far out of the cold water as floatplanes did. The Type CC did not serve very long in the German Navy and the aircraft were soon placed in storage.

However, Camillo Castiglione, head of the Austro-Hungarian branch of Brandenburg and for whom the Type CC was named, was aware that the Austro-Hungarian Navy needed a fighter, and Castiglione gave them a Type CC powered by a 185 hp Austro-Daimler engine that was assigned serial number A.12. This aircraft was presented to naval ace Gottfried Banfield, CO of the Trieste Naval Air Station, who stated that it was the best single-seat naval fighter so far. The Austro-Hungarian Navy then purchased a dozen more Type CC aircraft, serials A.13–A.24. These aircraft were supposed to be powered by 185 hp Hiero engines, although engine shortages meant the first four used the 160 hp Hiero.

A second batch of two dozen aircraft, A.25–A.48, to be powered by the 200 Hiero, was soon ordered. From A.31 on the aircraft featured a number of improvements, most notably mounting two fixed Schwarzlose machine guns and using an airfoil radiator instead of the previous car-type radiator. During production the fuselage was also lengthened for better directional stability. During the triplane craze one aircraft, A.45, was tested with a third wing mounted between the existing wings; climb was slightly improved but the additional weight and drag made it noticeably slower. The CC served successfully in the warmer waters of the Adriatic until replaced by the improved W.18.

Brandenburg CC Production		
Marine Numbers	**Qty**	**Notes**
946	1	Prototype, 1 gun, frontal radiator
1137–1146	10	Two guns, airfoil radiator
1327–1351	25	Two guns, airfoil radiator

Below: These four photos of the Brandenburg CC prototype show the early configuration of a single gun and frontal radiator for its 150 hp Benz Bz.III engine. The wood hull is stained and the flying surfaces are clear-doped linen.

Below: This CC from the first production batch displays a number of changes from the prototype; it has two guns, an airfoil radiator, and the engine is enclosed in a streamlined cowling.

Above: Brandenburg CC #1144 was from the first production batch. It carries two guns, has an airfoil radiator, and a streamlined engine cowling.

Above: Brandenburg CC from the first production batch, probably #1144, displays its clean lines. It carries two guns, has an airfoil radiator, and a streamlined engine cowling with propeller spinner.

Above: Brandenburg CC from the first production batch, probably #1144, displays its clean lines. It carries two guns, has an airfoil radiator, and a streamlined engine cowling with propeller spinner.

Above: Damaged Brandenburg CC #1331 was built in the second and final production batch. Two guns are carried and an airfoil radiator is fitted, as in the first production aircraft. The streamlined engine cowling is not fitted to this aircraft and may have been omitted from the entire series.

Above & Below: Like the KDW, later production Brandenburg CC fighters were fitted with additional interplane struts to stiffen the upper wing for improved aileron effectiveness. Two guns and an airfoil radiator are fitted.

Brandenburg W17

Above: Only two W17 fighters were built and one, serial A.49, was supplied to the Austro-Hungarian navy, where it was flown operationally by *Oblt*. Gottfried Banfield, the leading Austro-Hungarian Naval ace.

The Brandenburg W17 flying boat fighter was a development of the Type CC; the key difference was the use of a different wing cellule. The W17 had a smaller lower wing and the star-struts were replaced by slanted, parallel struts. One aircraft was tested by the German Navy and the other, A.49, shown above, was flown operationally by *Oblt*. Gottfried Banfield, the leading Austro-Hungarian naval ace, in late summer 1917, but no further production was undertaken. Mounting two fixed machine guns, the W17 was a transitional design between the production CC and the production W18.

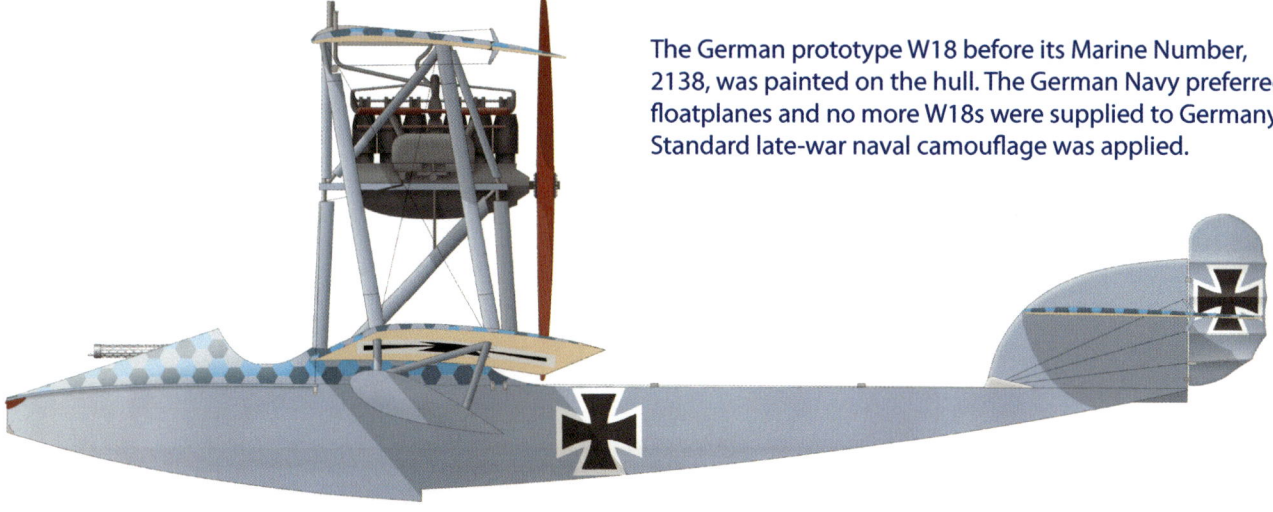

The German prototype W18 before its Marine Number, 2138, was painted on the hull. The German Navy preferred floatplanes and no more W18s were supplied to Germany. Standard late-war naval camouflage was applied.

Brandenburg W18

Above: The German Navy's sole W18; its Marine Number 2138 is chalked on the side of the fuselage.

The Brandenburg W18 flying boat fighter was a further development of the Type CC, this time using a more conventional wing cellule. One example, Marine #2138, was supplied to the German Navy, but with little German interest in flying boats only the single W18 was delivered.

However, like the Type CC, the W18 was widely used by the Austro-Hungarian Navy. In December 1916 47 aircraft, serials A.50–A.96, were ordered. These mounted two fixed Schwarzlose machine guns and were powered by the 230 hp Hiero engine, although some aircraft were delivered with 200 hp Hiero engines due to shortages of the more powerful engine. The W18s were used for both station defense and for escorting bombing raids on Italian targets.

Below: The W18 had conventional wing struts, two guns, and the standard late-war German naval camouflage.

Above: The German W18 had conventional wing struts, two guns, and the standard late-war German naval camouflage.

Below: The W18 fighters supplied to the Austro-Hungarian Navy had conventional wing struts, two guns, and plain finish with Austro-Hungarian markings; this is A.78.

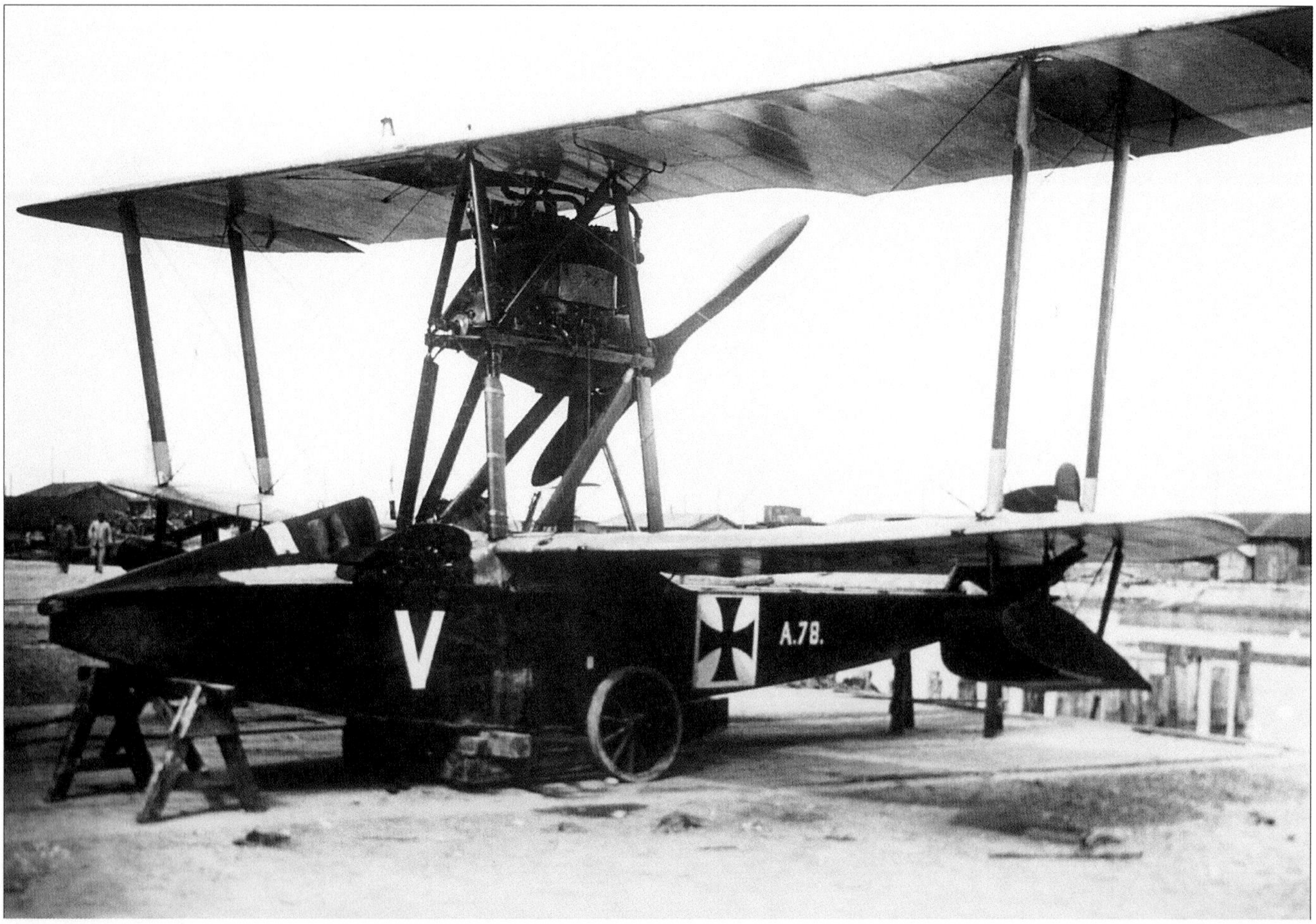

Above: A.71 was one of 47 Brandenburg W.18s that served with the Austro-Hungarian Navy, which liked flying boats. Fitted with two guns and powered by a 230 hp Hiero engine, the W.18 served successfully in the Adriatic until the end of the war. The red/white/red Austro-Hungarian colors can be seen under the upper right wingtip.

Below: This view of A.78. shows the red/white/red Austro-Hungarian markings under the upper wings and the iron cross insignia below the lower wings. Flying surfaces were clear-doped linen and the wood fuselage was stained.

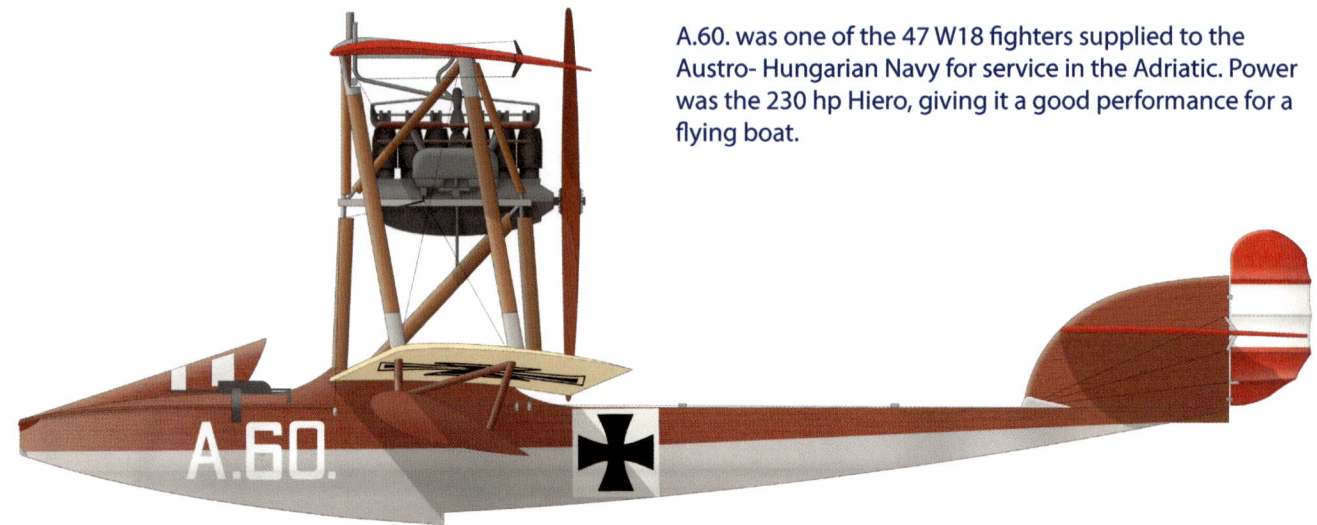

A.60. was one of the 47 W18 fighters supplied to the Austro- Hungarian Navy for service in the Adriatic. Power was the 230 hp Hiero, giving it a good performance for a flying boat.

Left: An Austro-Hungarian naval fighter pilot readies for his next mission in W18 A.50.

Below: W18 A.91 is somewhat the worse for wear after being captured by the Italians.

Two-Seat Biplane Fighters

Above: The combat effectiveness of the new Brandenburg W12 two-seat fighter immediately rendered all other floatplane fighters obsolete. All subsequent production floatplane fighters were developed from the W12. Here a pair of W12 fighters escort a U-Boat into harbor on the Flanders coast.

The German Navy's need for a seaplane station defense fighter led in 1916 to development of a number of single-seat seaplane fighters. As the air war intensified, the German Navy needed aircraft that could carry out longer-range, more offensive operations, especially fighter sweeps over the North Sea to counter the large British Curtiss and Felixstowe flying boats that were conducting anti-submarine operations against German U-boats. Two-seat fighters were preferred for greater combat effectiveness; a key additional benefit was the observer/gunner could assist the pilot with over-water navigation, a serious challenge when electronic navigation aids were not yet available. In addition, the observer had a flexible gun for greater combat effectiveness, and some aircraft carried wireless operated by the observer, a great advantage over water that single-seaters could not use.

Prototypes to fulfill the Navy's requirement were produced by four manufacturers, Albatros, Brandenburg, Friedrichshafen, and Sablatnig. Albatros, the largest German aircraft manufacturer, submitted the Albatros W8. Friedrichshafen,

the largest German manufacturer of floatplanes, submitted the Friedrichshafen FF63. Sablatnig, a minor manufacturer, submitted the SF3 and SF7 designs. The Naval Shipyard at Wilhelmshafen, which was not actually a manufacturer but built a small number of prototype seaplanes, none produced in quantity, submitted the K.W. (Wilhelmshafen) No.945. Finally Brandenburg, dependent on sales to the German Navy because the Army refused their designs after a series of structural failures during testing early in the war, submitted their W12.

The design chosen for production was the excellent Brandenburg W12 biplane. Introduced into service in the fall of 1917, the W12 immediately proved far more effective than its single-seat predecessors. The highly successful W12 was further developed to improve its operational effectiveness. One direction this development took was increased range and endurance that were provided by a larger, more powerful aircraft of similar configuration, the Brandenburg W19. The other direction was greater speed, which resulted in monoplane developments discussed in the next chapter.

Two-Seat Biplane Floatplane Fighter Production

Aircraft	Quantity	Marine Numbers
Albatros W8	3	5001–5003
Brandenburg W12	146	1011–1016, 1178–1187, 1395–1414, 2000–2019, 2023–2052, 2093–2132, 2217–2236
Brandenburg W19	115	1469–1471, 2207–2216, 2237–2257, 2259–2278, 2544–2563
Brandenburg W27	3	2201–2203
Brandenburg W32	3	2282–2284
Friedrichshafen FF48	3	1472–1474
Sablatnig SF3	1	619
Sablatnig SF7	3	1475–1477
K.W. No.945	1	945

Specifications for Two-Seat Biplane Floatplane Fighters

Type	Albatros W8	Brandenburg W12	Brandenburg W19	Friedrichshafen FF48	Sablatnig SF7
Engine	195 hp Benz Bz.IIIb	150 hp Benz Bz.III or 160 hp Mercedes D.III	240 hp or 260 hp Maybach Mb.IVa	240 hp Maybach Mb.IVa	240 hp Maybach Mb.IVa
Span, Upper	11.46 m	11.20 m	13.80 m	16.25 m	—
Length	9.59 m	9.6 m	10.65 m	11.2 m	—
Empty Weight	—	997 kg.	1,435 kg.	1,591 kg	—
Loaded Weight	—	1,454 kg.	2,005 kg.	2,216 kg	2,120 kg
Maximum Speed	150 km/h	160 km/h	151 km/h	153 km/h	162 km/h
Climb to 1,000m	6.5 min.	7 min.	6.4 min.	6.4 min.	8 min.
Climb to 2,000m	—	18.9 min.	18.9 min.	—	—
Climb to 3,000m	34 min.	—	23 min.	—	36 min.
Flight Duration	3.5 hours	3.5 hours	5 hours	5.75 hours	—
Armament	1–2 fixed guns + 1 flexible gun	1 flexible gun + 1 or 2 fixed machine guns	2 fixed guns + 1 flexible gun or 20mm cannon	1 flexible gun + 1 fixed gun	1 flexible gun + 1 fixed gun

Brandenburg W12

Above: Marine #1407, a Brandenburg W12 of the second production batch, illustrates the distinctive features of this breakthrough design, including the upswept tail without vertical fin that gave the gunner an excellent field of fire. The extensive keel surface provided by the deep rear fuselage eliminated the need for a fin. To reduce drag the radiator has been moved to the nose. Both doors on the pilot's cockpit are open to allow easier access.

The Brandenburg W12 two-seat floatplane fighter was a breakthrough design. Despite being a two-seater using the same engines powering the smaller single-seaters it replaced, it had similar speed and greater range coupled with better maneuverability and flying characteristics!

Its two-seat configuration also provided greatly improved air-to-air combat effectiveness from its combination of fixed and flexible armament, and the second crewman was able to assist the pilot with over-water navigation, a particular operational challenge in the days before electronic navigation aids. Furthermore, some W12s carried wireless senders and receivers for the observer, something none of the single-seat fighters could do, which was a tactical advantage. The W12 made all preceding single-seat seaplane fighters obsolete at a stroke.

The key secret of its success was its innovative structural design that used its sturdy float bracing to also support the wings, eliminating the need for separate, drag-producing bracing wires, a key to its good speed. Furthermore, its innovative tail design gave the observer excellent visibility and field of fire. In addition, the observer and his gun ring were mounted high enough that he could fire forward over the upper wing, giving him an unexcelled field of fire and further enhancing combat effectiveness.

Despite its general excellence, the basic design was subject to a great deal of fine tuning to improve stability, maneuverability, and sea handling. In fact, none of the Brandenburg two-seat fighters were ever able to handle sea states as rough as the robust Friedrichshafen reconnaissance floatplanes, and there were continual problems with float maintenance. Fuselage length was extended in later aircraft to improve longitudinal stability, and late production aircraft had ailerons on all four wings for improved maneuverability. The center section was also redesigned during production to improve the pilot's field of view and ease of egress in emergencies. Some 30 later aircraft were fitted with two fixed guns for the pilot, and some aircraft with only a single gun for the pilot were fitted with wireless.

Brandenburg W12 Production Orders

Order Date	Marine Numbers	Qty	Class	Engine	Notes
15 Oct. 1916	1014–1016	3	C2MG	Mercedes D.III	1014 was prototype, short fuselage
22 Nov. 1916	1011–1013	3	C2MG	Mercedes D.III	Short fuselage
5 Jan. 1917	1178–1187	10	C2MG	Benz Bz.III	1185 had longer fuselage
13 Mar. 1917	1395–1414	20	C2MG	Benz Bz.III	Short fuselages; 1413 had four ailerons
10 Sep. 1917	2000–2019	20	C2MG	Benz Bz.III	Long fuselages start with this series. Larger wing cut-out.
Oct. 1917	2023–2052	30	C3MG	Benz Bz.III	Two fixed machine guns. 2027 destroyed during acceptance testing.
Oct. 1917	2093–2112	20	C2MG	Mercedes D.III	—
Oct. 1917	2113–2132	20	C3MG	Mercedes D.III	Two fixed machine guns
Nov. 1917	2217–2236	20	C2MGHFT	Mercedes D.IIIa	Wireless equipment fitted

Note: Of 146 W12 aircraft built, 96 aircraft had one fixed gun and 50 aircraft had two fixed guns.

Below: W12 Marine #1012, the second pre-production aircraft, is seen here in operational markings at Zeebrugge. As tabulated above, the pre-production aircraft were category C2MG and were powered by the 160 hp Mercedes D.III engine. This colorful aircraft is depicted in the cover painting and also in the painting with the U-Boat.

Above: W12s at their base at Zeebrugge. *Oblt.z.S.* Christiansen's #1183 is in the center.

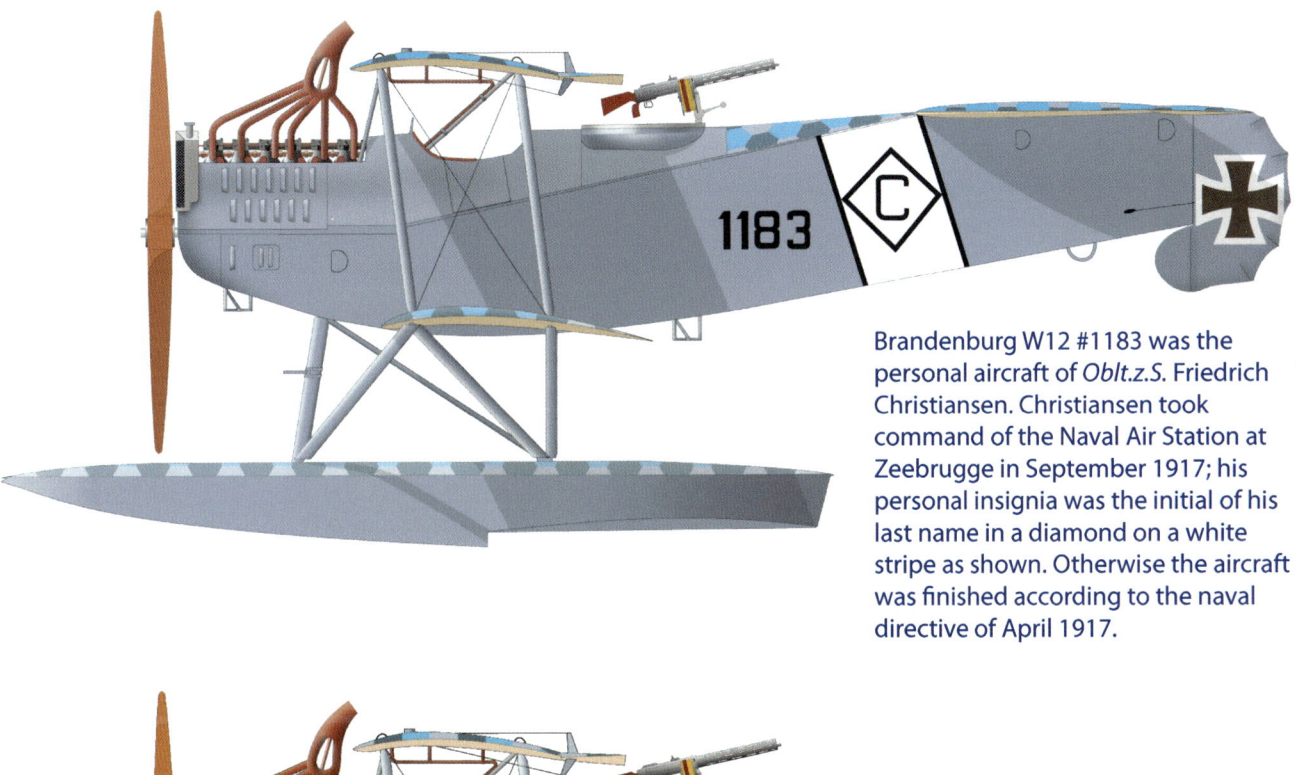

Brandenburg W12 #1183 was the personal aircraft of *Oblt.z.S.* Friedrich Christiansen. Christiansen took command of the Naval Air Station at Zeebrugge in September 1917; his personal insignia was the initial of his last name in a diamond on a white stripe as shown. Otherwise the aircraft was finished according to the naval directive of April 1917.

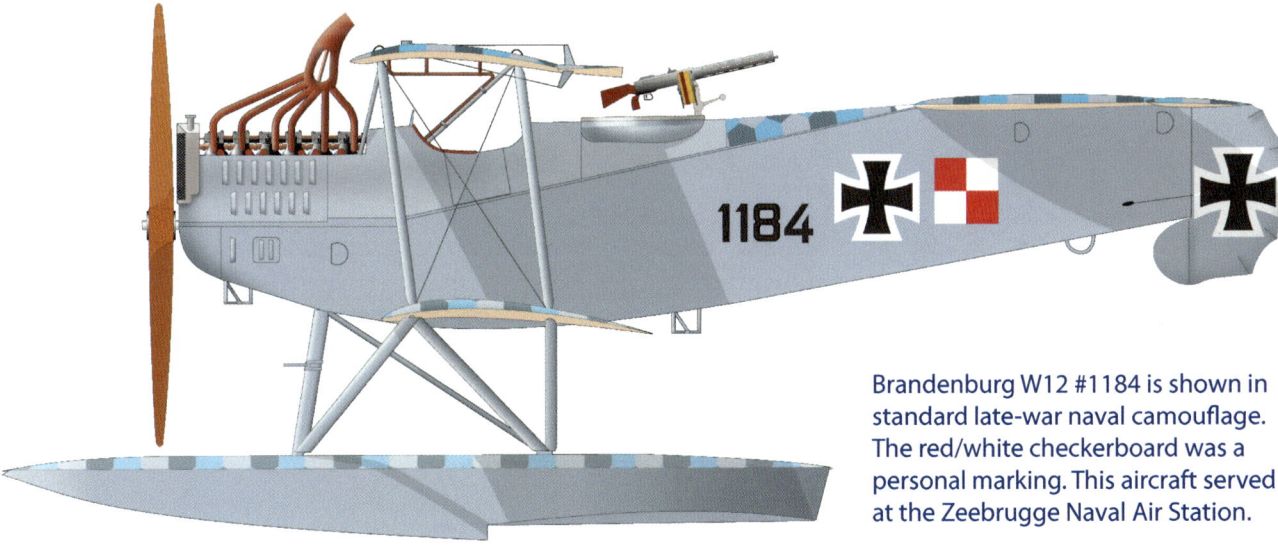

Brandenburg W12 #1184 is shown in standard late-war naval camouflage. The red/white checkerboard was a personal marking. This aircraft served at the Zeebrugge Naval Air Station.

Above: Marine #1409 is first in this lineup of W12 fighters at Zeebrugge Naval Air Station on the Flanders coast.

Below: Brandenburg W12 #1409 was accepted in the last half of November 1917 and written off on May 6, 1918. The crew's personal insignia is on the side of the observer's cockpit. Unfortunately, their names are not known.

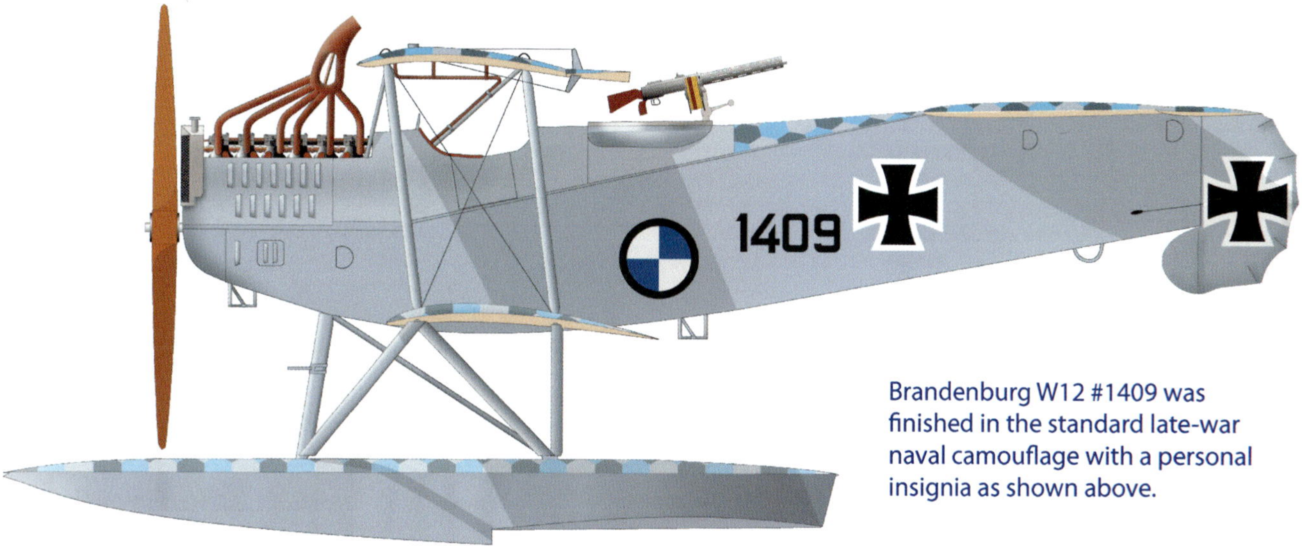

Brandenburg W12 #1409 was finished in the standard late-war naval camouflage with a personal insignia as shown above.

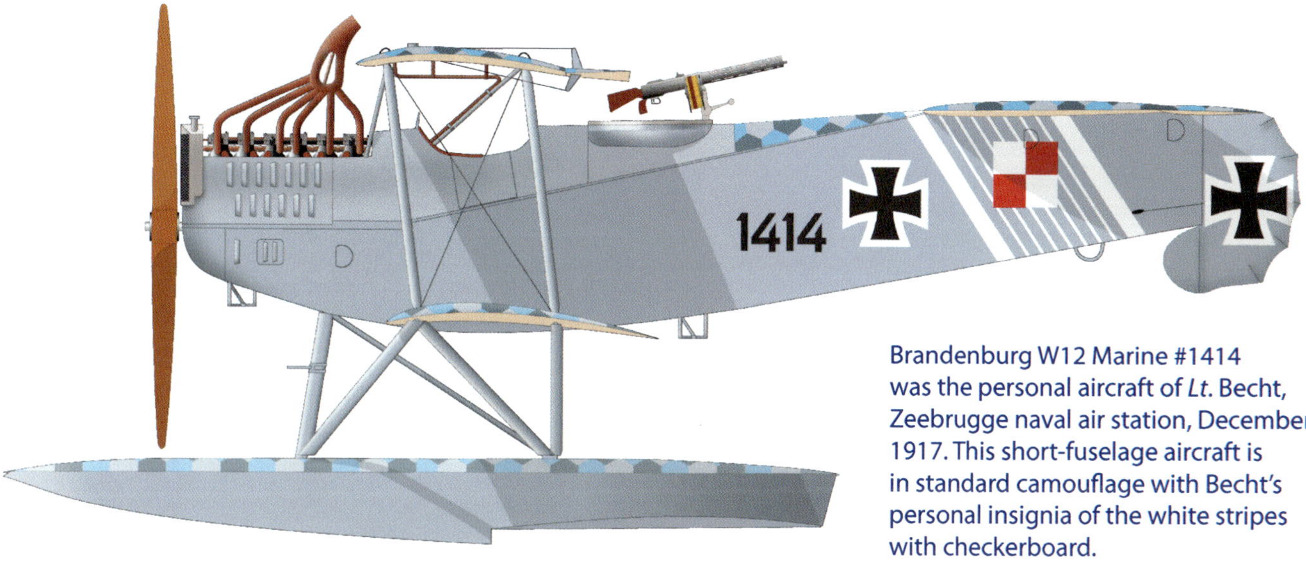

Brandenburg W12 Marine #1414 was the personal aircraft of *Lt.* Becht, Zeebrugge naval air station, December 1917. This short-fuselage aircraft is in standard camouflage with Becht's personal insignia of the white stripes with checkerboard.

Brandenburg W12 Marine #2002 is from the first series with the longer fuselage that improved stability. It is in standard camouflage with Bremen's coat of arms (white key on red) as a personal insignia. This aircraft had ailerons on all wings.

Above: W12 Marine #1407 of the second production batch in an embarrassing position. Its original insignia have been over-painted in the new insignia standardized on March 30, 1918. The single white stripe on the rear fuselage indicates assignment to a specific naval air station, possibly Borkum.

Above: W12 Marine #1410, accepted during the second half of November 1917, wears an interesting personal insignia. It was photographed with several crew members on February 6, 1918. It was written off on April 23, 1918.

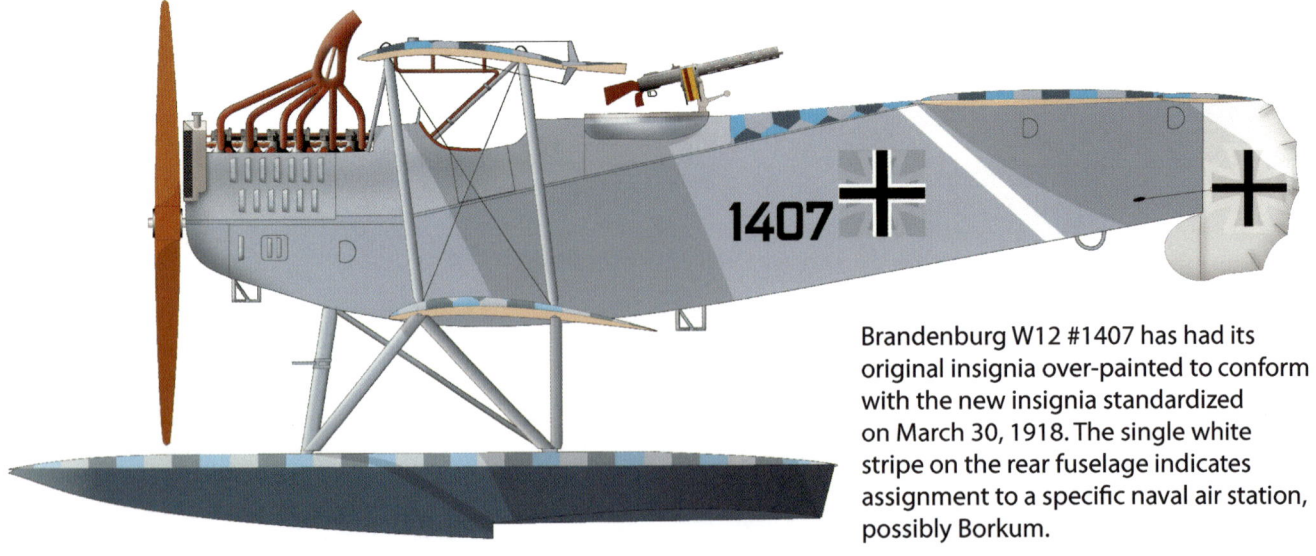

Brandenburg W12 #1407 has had its original insignia over-painted to conform with the new insignia standardized on March 30, 1918. The single white stripe on the rear fuselage indicates assignment to a specific naval air station, possibly Borkum.

Above: W12 #2004, a category C2MG, has the longer fuselage plus a nose radiator for its Benz Bz.III. The men in the group photo are the pilots of *Seeflugstation Flanders I*, abbreviated *See I*, in late 1917.

Below: The four-leaf clover on the fuselage worked; this crew is happy to survive a forced landing on a grassy field. The W12 was a sturdy warplane and this aircraft turned out to be almost undamaged by its experience.

Above: Marine #2052, a Brandenburg W12 of the fourth production batch, has a nose radiator and ailerons on all wings, with upper and lower ailerons connected by an actuating strut. It also mounted two fixed machine guns for the pilot in addition to the observer's flexible gun, making it a category C3MG. The outer portion of the propeller leading edge was covered with metal to reduce erosion from water spray.

Above: Brandenburg W12 #2108 was a category C2MG. Some German army air service observers were given short courses in over-water navigation, accounting for their presence in the photograph.

Above: W12 #2128, a C3MG, in difficulty.

Above: Brandenburg W12 Marine #2094 has come to grief between the Helgoland launching ramps. *Staffel* number 3 is just behind the interim national insignia.

Facing Page: *Oblt.z.S.* Friedrich Christainsen downed British airship *C27*; that victory lead to his award of the *Pour le Mérite*. He is shown here with his gunner, *Vzfw.* Wladika, in W12 #1183, the aircraft he used to destroy *C27*. W12 #1183 was accepted in September 1917 and served until destroyed in a bombing raid on the Zeebrugge Mole on May 10, 1918.

Brandenburg W19

Above: This unmarked W19 was certainly the first prototype. A key indication this was the first prototype W19 is the lack of ailerons on the lower wing; all other W19s had ailerons on both upper and lower wings for improved maneuverability. The horn exhaust extends horizontally from the engine cowling.

Nearly identical to the W12 in configuration, the W19 was considerably enlarged to enable it to carry the greater fuel load needed for the longer range and endurance desired. Other than its larger size, the main visible difference was the W19 had two-bay interplane struts to support its larger wings instead of the single-bay struts of the smaller W12. Like late production W12 aircraft, all W19 aircraft except the first prototype had ailerons on all four wingtips for enhanced maneuverability. With the exception of the first three W19 prototypes, which had a single fixed machine-gun for the pilot, all subsequent W19 production aircraft had two fixed machine guns.

One early W19, Marine #2237 ordered in

Brandenburg W19 Production				
Series	**Marine Numbers**	**Qty**	**Category**	**Armament & Notes**
Prototypes	1469–1471	3	C2MG	1 fixed, 1 flexible MG
Series 1	2207–2216	10	C3MG	2 fixed, 1 flexible MG
Series 2	2237	1	CK	2 fixed MG, 1 flexible 20mm cannon
	2238–2257	20	C3MG	2 fixed, 1 flexible MG
Series 3	2259–2275	17	C2MGHTF	1 fixed, 1 flexible MG, wireless xmtr/rcvr
	2276–2278	3	CK	2 fixed MG, 1 flexible 20mm cannon
	2537	1	C3MG	2 fixed, 1 flexible MG
Series 4	2544–2563	20	CK	2 fixed MG, 1 flexible 20mm cannon
Series 5	2683–2722	40	C3MG	2 fixed, 1 flexible MG

Above: Marine Number 1470 was the second prototype W19. The aircraft is painted and marked and the engine cowling appears to be designed for individual exhaust pipes. The float bracing appears very sturdy. Only the over-wing radiator spoils its clean lines.

December 1917, had a flexible Becker 20mm cannon instead of a Parabellum for the observer. This seaplane apparently had an enlarged rudder and elevator, perhaps for greater stability with the cannon in the slipstream, although further details are lacking. After testing in April 1918, during which fifty rounds were rapid-fired with no installation problems, the only change requested was enlarging the observer's gun ring from 900mm to 1000mm diameter. After approval by front-line personnel, #2237 was sent to the front for evaluation, where it was written off on 20 June. Apparently the evaluation was successful because in early June the Navy ordered the fourth W19 production series,

Marine Numbers 2544–2563, armed with the Becker cannon. All 20 W19s of this series were found at Warnemünde in December 1918 by the Allied Naval Armistice Commission, most still in their packing crates.

Most W19 aircraft were powered by the 240 hp Maybach Mb.IVa, making the W19 much more powerful than the W12. This significant power increase compensated for its greater size and weight, making the W19 nearly as fast as the W12 while carrying more fuel, armament, and equipment. In April 1918 the first W19 arrived at Zeebrugge to supplement the W12 in the North Sea.

W19 Production Notes:

1. Marine Number 1469 was destroyed on its first flight in August 1917.
2. The engine for #1469 is not confirmed; #1470 & #1471 had the 240 hp Maybach Mb.IVa.
3. The engine for production series 1, 2, and 3 was usually the 240 hp Maybach Mb.IVa; some aircraft had the 260 hp Maybach Mb.IVa.
4. Marine Numbers 2208–2216 and 2237–2239 were accepted the second half of April/first half of May 1918.
5. Marine Number 2237, accepted in April 1918, was class CK; this was a test installation of the 20mm Becker in the W19.
6. Marine Number 2267, accepted in the second half of June 1918, had a 260 hp Maybach Mb.IVa.
7. The engine for production series 4 was the 260 hp Maybach Mb.IVa.
8. Completion of production series 5 is not confirmed, but #2687 and #2688 were handed over to Italy on 15 September 1920, so some aircraft of this series were produced.
9. A total of 115 W19s were ordered but it is not known if all of production series 5 were delivered. Production of at least 77 W19s is confirmed.

Above: This side view of Marine Number 1470, the second prototype W19, emphasizes its clean lines for a biplane. An actuating strut connects the ailerons on upper and lower wings.

The W19 and the Maybach Mb.IVa

All W19s used the Maybach Mb.IVa. W19s through the first three production batches were generally scheduled to receive the 240 hp Maybach Mb.IVa engine, but when struck off charge some individual airplanes had the high-compression 260 hp Maybach Mb.IVa engine. It is not known if the more powerful engine was installed during production or retrofitted after delivery. The fourth production series was scheduled to have the 260 hp Maybach Mb.IVa. On 12 August 1918 this series was allotted between *Kofl F* (North Sea) and *Kofl Marinekorps* (Flanders).

Above: This front view shows the very clean lines of the W19 despite its two-bay wing bracing. The shape of the exhaust manifold is the same as in the photo of the unmarked W19. This could be that same aircraft after modification; unlike that aircraft, it has ailerons on both upper and lower wings. Alternatively, it might be the third prototype. The lack of a second fixed gun for the pilot on the port side is a key indication this aircraft is one of the first three prototypes.

Above: The clean appearance and small spinner are shown in this view of Marine Number 2207, the first production aircraft. Compared to earlier airframes the exhaust manifold has been lengthened and angled nearly vertically to exhaust the gases over the wing, away from the crew.

Confusingly, both the Maybach 240 hp and 260 hp engines shared the designation Mb.IVa. The engine rated at 240 hp had cast iron pistons, produced a maximum of 245 hp, and weighed 400 kg. The engine rated at 260 hp had aluminum pistons, produced a maximum of 300 hp, and weighed 390 kg. Both engines had six cylinders of 165mm bore and 180mm stroke, providing 23.1 liters capacity and developed their rated power at 1,400 RPM. From these similarities it is apparent they were essentially the same engine with different pistons, power ratings, and compression ratios, but with little or nothing to distinguish between them externally.

Above: This view of Marine Number 2207 emphasizes its streamlined fuselage and the robust float bracing. The fairing over the pilot's left gun shows dedication to streamlining and protecting the gun mechanism from the airflow and especially salt-water spray. The three-color naval lozenge fabric covering the upper surfaces of the wings and fuselage are just visible; this camouflage fabric was also used on the top of the tailplane.

Above: This side view of Marine Number 2207 emphasizes the gunner's clear field of fire. The actuating strut between the ailerons on upper and lower wings has now been streamlined to reduce drag.

Above: Front view, probably of Marine Number 2207, showing its over-wing radiator and twin forward-firing machine guns. The nearly vertical exhaust manifold is different than that of the aircraft in the other front view.

BENZ
Automobile und Flugmotoren

Benz & Cie. Rheinische Automobil- und Motoren-Fabrik A.-G., Mannheim

Left: WWI advertisement for the Benz company, one of the largest German aero engine manufacturers.

Right: W19 Marine Number 2237 photographed at Warnemünde on 25 April 1918 during ground firing of the 20mm Becker cannon. The gunner appears somewhat cramped, which led to the recommendation to enlarge the gun ring diameter from 900mm to 1000mm.

Below: Another view of the 20mm Becker mounted in W19 Marine Number 2237 on 25 April 1918 showing details of the gun ring.

Above: A modified 20mm Becker cannon mounted in W19 Marine Number 2237 on 25 April 1918. The cannon now has a shoulder stock and a bag to catch empty shells.

Facing Page: This view of the 20mm Becker cannon mounted in W19 Marine Number 2237 shows the gun elevation available to the observer. The hexagonal naval camouflage is clearly visible.

Below: A W19 from Norderney, showing its white chevron unit marking on the rear fuselage, takes off on another mission.

Brandenburg W27

Above: Marine #2202 was the second Brandenburg W27 prototype. The W27 was a development of the W12 with streamlined struts and a 195 hp Benz Bz.IIIb V-8 engine. One drawback of the struts was that they blocked some of the crews' field of view. The engine never reached full production and the W27 was limited to only three prototypes.

The W27 differed from the W12 primarily by its use of I-type interplane struts and a 195 hp Benz Bz.IIIb V-8 engine. These were class C3MG, so armament was two fixed Spandau machine guns for the pilot and one flexible Parabellum machine gun for the observer. Three aircraft, Marine Numbers 2201–2203, were built, but the engine was not in full production and the aircraft were used as trainers.

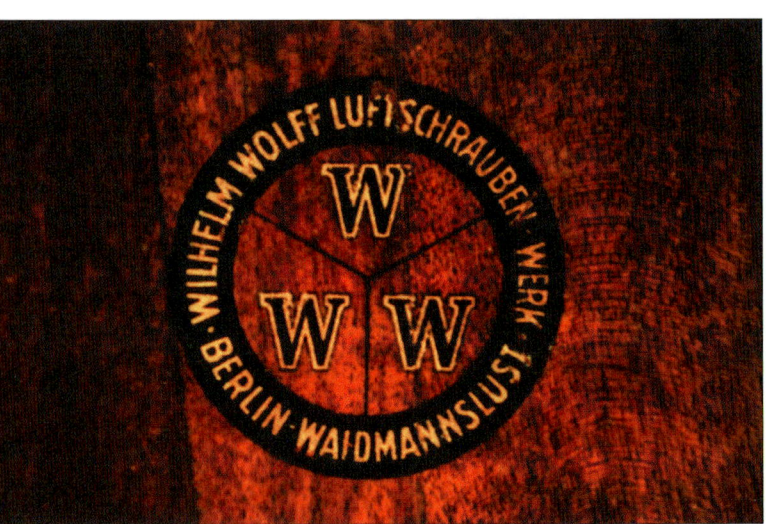

Above: The logo of the German Wolff propeller company on a Wolff propeller.

Above: WWI advertisement for Mercedes engines for automobiles and airplanes. Mercedes was the trade name of the products produced by the Daimler company, the largest producer of German aero engines in WWI, followed closely by Benz.

Brandenburg W32

Above & Below: Marine #2282 was the first Brandenburg W32 prototype. The W32 was a development of the W27 with a 170 hp Mercedes D.IIIa engine. The aircraft was no improvement over the W12 and was limited to three prototypes.

Yet another W12 derivative was the W32, which appears to have been a W27 fitted with a 170 hp Mercedes D.IIIa engine. Again three aircraft were apparently built; Marine Numbers 2282–2284. The first aircraft, #2282, was class C3MG; the other two were class C2MG HFT. Dimensions were the same as the W27 and the W32s were accepted about the end of June 1918. By this time the faster W29 and W33 monoplanes were in production and an improved W12 was unnecessary.

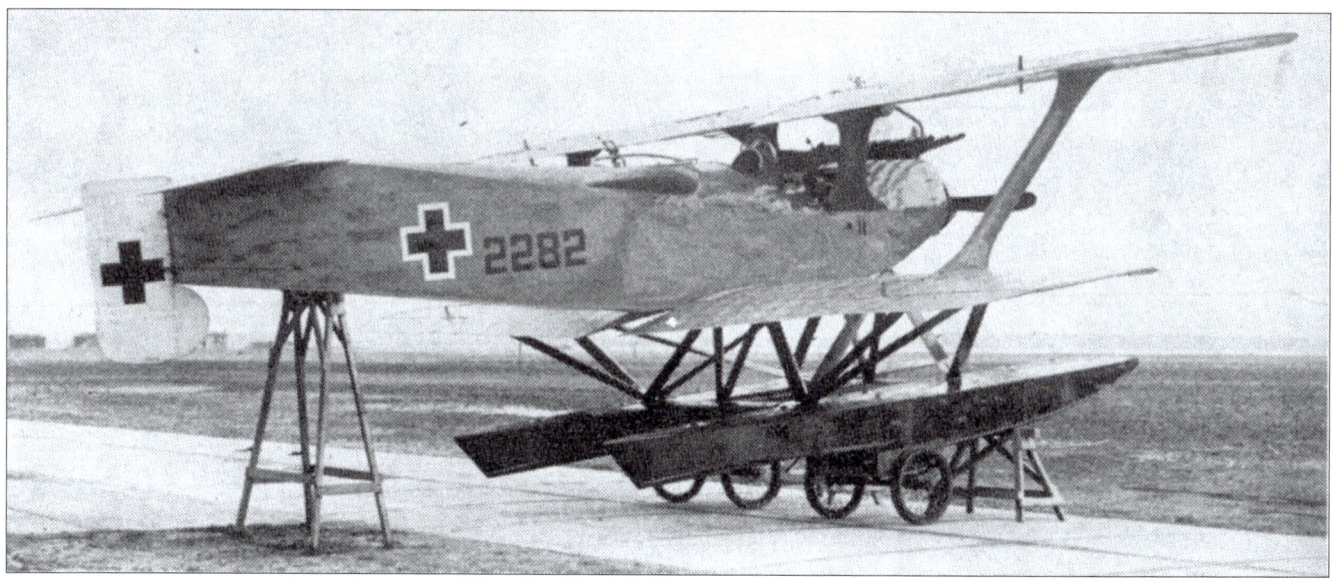

Albatros W8

Above: The first prototype Albatros W8 was a handsome, well-streamlined two-seat fighter powered by a V-8 that was not yet in mass production. However, the flat radiator under the wing somewhat spoiled the overall streamlining.

The Albatros company, largest aircraft manufacturer in WWI Germany, offered the Albatros W8 for the Navy's two-seat fighter. Although similar in concept to the Brandenburg W12, its tail design obstructed more of the gunner's field of fire and it used an experimental engine that reached production too late to power aircraft at the front; only three were built.

Above & Left: Two views of Marine #5002, the second Albatros W8 prototype, shows that the elegant spinner of the prototype was abandoned. Otherwise #5502 appears to incorporate no other visible changes.

Friedrichshafen FF48

Above: The Friedrichshafen FF48 was a larger aircraft in the class of the W19 and like the W19 was powered by the 240 hp Maybach Mb.IVa.

Friedrichshafen, who produced more seaplanes than any other German company during the war, built their FF48 to the same requirement as the W19. The FF48 had good speed and climb for such a large floatplane but may have lacked the necessary maneuverability, and only three were built, indicating it had no significant advantages over the W19. Certainly it required more wire bracing than the W19 and did not have the look of a fighter.

Above: WWI advertisement for Mercedes engines for automobiles and airplanes.

Right: The logo of the German Integral propeller company.

Sablatnig SF3

Above & Below: The Sablatnig SF3 was compact for a two-seat fighter. However, the SF3 featured a profusion of struts and bracing wires that certainly created more drag than the cleverly-designed W12 and only one prototype was built.

Sablatnig, a small firm that specialized in seaplanes, produced two two-seat floatplane fighter designs. First was the SF3, a sturdy-looking aircraft powered by a 220 hp Benz Bz.IV. The drag of the streamlined fuselage was more than compensated for by its multitude of struts and bracing wires, and it remained a single prototype.

Sablatnig SF7

Above: The SF7 was a massive aircraft, looking more like a typical reconnaissance floatplane than a two-seat fighter. It probably lacked maneuverability competitive with that of the W19. Marine #1475 was the first of three prototypes.

The SF7, powered by a 240 hp Maybach Mb.IVa, was the second Sablatnig design for a two-seat naval fighter. It had good speed but despite that only three, Marine Numbers 1475–1477, were built.

Below: The Sablatnig SF7 was in the W19 class and was powered by the same engine. Although somewhat faster than the W19, only three were produced. The I-struts appear to interfere with the crews' field of view.

K.W. (Wilhelmshafen) No.945

Above: The sole prototype of the K.W. (Wilhemlshafen) No.945 built was clearly inspired by the Brandenburg W12, but lacked the qualities to gain a production order.

The Naval Shipyard at Wilhelmshafen built a small number of prototype seaplanes, none of which were produced in quantity. One prototype built there was the K.W. No.945, a two-seat naval fighter clearly inspired by the Brandenburg W12. It was powered by a 150 hp Benz Bz.III engine, a powerplant used by many Brandenburg W12s. Other than its engine, few details survive.

Above: The logo of the Germania propeller company on a Germania propeller.

Above: A WWI advertisement from the Benz company, one of the largest manufacturers of German aero engines.

Two-Seat Monoplane Fighters

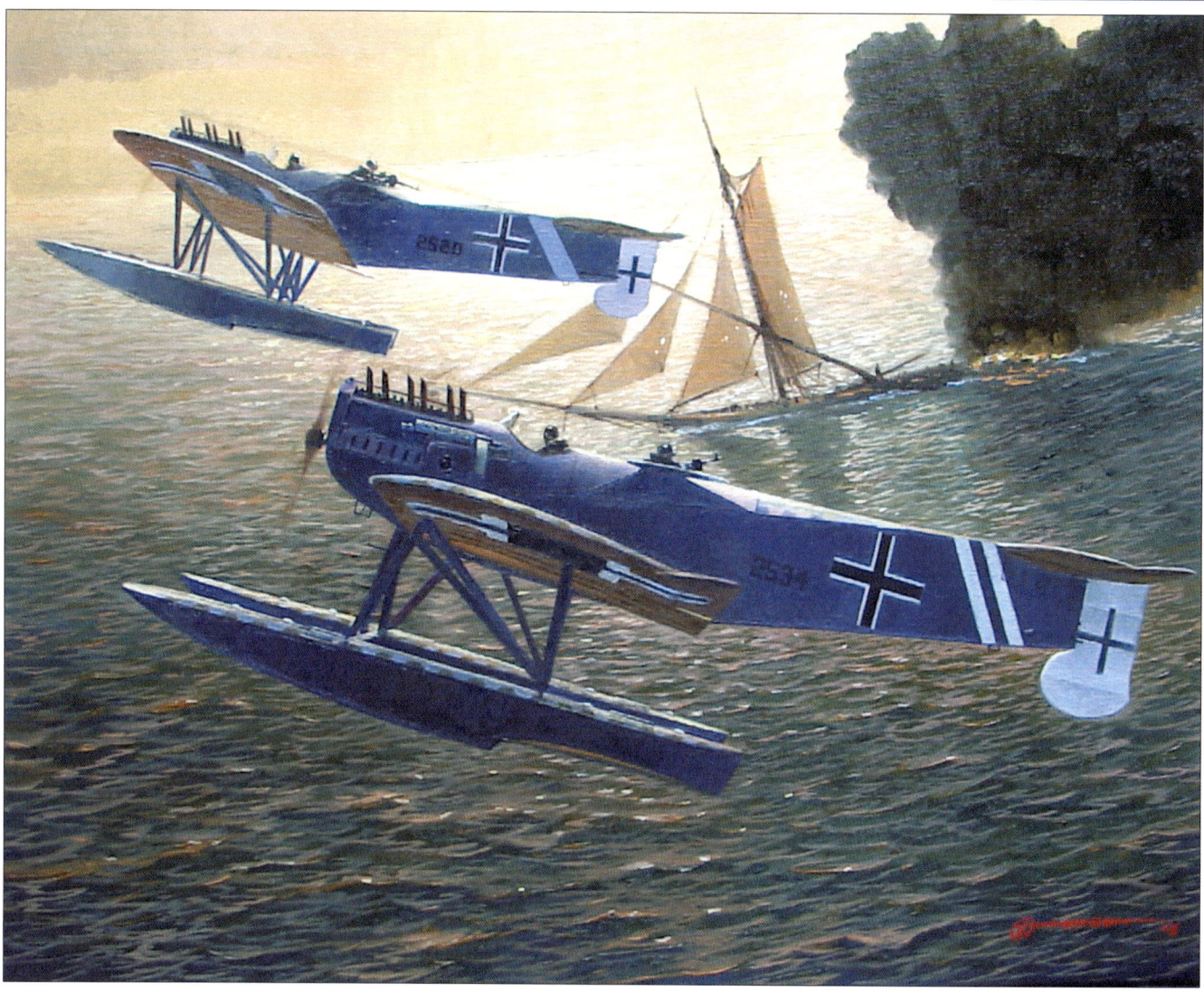

Above: A pair of Brandenburg W29s sink a sloop carrying contraband to Holland.

To extend its range and endurance, qualities especially valuable over the North Sea, the highly successful Brandenburg W12 biplane design was enlarged to carry more fuel, resulting in the W19.

Similarly, more speed was desired. Although the W12 was fast enough to intercept the formidable Felixstowe flying boats, the closing speed was not as great as wanted. This sometimes resulted in long tail chases, and there was a concern that W12s might have to abandon pursuit due to low fuel state.

The basic W12 biplane configuration was developed into a monoplane for reduced drag and greater speed in the form of the Brandenburg W29. Finally, the W29 monoplane itself was followed by enlarged developments, the W33, W34, and W37, to combine its greater speed with the longer range and endurance of the Brandenburg W19.

The Brandenburg W29 went into operational service in April/May 1918 and had an immediate impact on the fighting over the North Sea. The larger W33 monoplane, built in small numbers, followed it into combat in late summer, and further enlarged developments were being pursued when the armistice terminated further work. Three W34 prototypes were actually completed and three W37 prototypes were being built, but construction was abandoned in December 1918. The W29, W33, and W37 saw extensive and successful post-war service in Norway, Denmark, Finland, and Sweden.

Two-Seat Monoplane Naval Fighters		
Aircraft	**Qty**	**Marine Numbers**
Brandenburg W29	209	2204–2206, 2287–2300, 2501–2506, 2507–2536, 2564–2583, 2584–2589, 2593–2682, 2730–2759, 2760–2789 (last 30 ordered not delivered)
Brandenburg W33	7	2538–2540, 2543, 2683–2684, 2726
Brandenburg W34	3	2727–2729
Brandenburg W37	0	2723–2725 ordered but canceled Dec. 1918 and not completed
Friedrichshafen FF63	1	?
Junkers CLS.I	3	7501–7503
Zeppelin-Lindau (Dornier) Cs.I	3	8501–8503 (only 8502 flown)

Above: W29 Marine #2532 from Norderney (indicated by the two identification stripes on the aft fuselage) on patrol over the North Sea. Like #2532, most W29s were category C3MG and had two fixed machine guns for the pilot. If a wireless transmitter and receiver were installed, the pilot had only one fixed gun to avoid over-loading the aircraft and reducing its performance. The Brandenburgs were at their most vulnerable when taking off and landing, and since they were based near the front there was always the possibility that they would be attacked by land-based fighters of the Royal Naval Air Service or, after the RNAS and RFC were combined on April 1, 1918, the former RNAS units now part of the RAF. That was why the naval aircrews adopted the tactic of taking off and landing in formation, and maintaining a stepped vee formation in flight with clear fields of fire for all the gunners. Without the handicap of floats the British fighters, mostly Sopwith Camels, had the advantages of speed, climb, and maneuverability, so formation flying for mutual defense was a priority for the crews.

Specifications for Two-Seat Monoplane Naval Fighters

Type	W29	W33	W34	W37	Junkers CLS.I	Dornier Cs.I
Engine	185 hp Benz Bz.IIIa	245 hp Maybach Mb.IVa	300 hp BuS.IVa	220 hp Benz Bz.IV	200 hp Benz Bz.IV	195 hp Benz Bz.IIIb
Span	13.5 m	15.85 m	16.60 m	17.70 m	12.75 m	13.18 m
Length	9.36 m	11.10 m	11.10 m	12.39 m	8.95 m	—
Wing Area	32.2 sq.m.	44.0 sq.m.	49.0 sq.m.	55.0 sq.m.	—	—
Empty Weight	1,000 kg	1,420 kg	1,534 kg	1,500 kg	914 kg	960 kg
Loaded Weight	1,494 kg	2,050 kg	2,270 kg	2,177 kg	1,420 kg	1,479 kg
Maximum Speed	168 km/h	173 km/h	175 km/h	155 km/h	180 km/h	150 km/h
Climb to 1,000m	6 min.	5.4 min.	—	—	—	—
Climb to 2,000m	13 min.	12.8 min.	—	—	—	—
Climb to 3,000m	23 min.	22.3 min.	—	—	—	—
Service Ceiling	5,000m (16,400 ft.)	—	—	—	—	—
Flight Duration	4 hours	—	—	—	—	—
Armament	1 flexible gun + 1 fixed machine gun (40) or 2 fixed machine guns (38)	1-2 fixed guns + 1 flexible machine gun or 1 20mm Becker cannon	1-2 fixed machine guns + 1 flexible machine gun	1-2 fixed machine guns + 1 flexible 20mm Becker cannon	2 fixed machine guns + 1 flexible machine gun	1 fixed machine gun + 1 flexible machine gun

Above: The prototype W29 #2204 with chief engineer Ernst Heinkel (in coat with fur collar third from right) with *Oblt.z.S.* Friedrich Christiansen on his right and Karl Heinkel and *Oberingenieur* Schweigert on his left.

Above: W29 in Norwegian service postwar. The W29 and W33 enjoyed long, successful postwar careers in Nordic countries.

Above: Closeup of a W29 nose showing the gun and engine details and overall aerodynamically clean design.

Brandenburg W29

Above: A Brandenburg W29 with two identification bands painted on its rear fuselage flies over a base, mostly likely Nordeney. The two bands signify Norderney as the naval air station to which the aircraft was assigned.

The W29 monoplane was based on the fuselage, engine, and floats of the W12 biplane. The sturdy struts bracing the floats did double duty bracing the wing, which had about the same area as both wings of the W12. The greatly reduced drag of the monoplane wing resulted in a notable speed increase with the same engines and payload. The increased speed of the W29 was welcomed because it made interception of the large Felixstowe flying boats easier, and the new monoplane set the standard for all future German wartime designs in its class.

W29 production aircraft were powered by the 150 hp Benz Bz.III, 185 hp Benz Bz.IIIa, 170 hp Mercedes D.IIIa, or 185 hp BMW.IIIa, and had one or two fixed machine guns for the pilot (depending on the series) and one flexible gun for the observer. Elimination of the upper wing not only made the monoplanes faster, but stability was improved, as was the observer's field of fire and pilot's field of view.

By naval standards the W29 was built in large numbers; 14 orders were submitted for a total of 239 aircraft, and 209 were built; the last batch of 30 ordered on 10 September 1918 was not completed due to the armistice.

Above: Ground crew ready a basket of carrier pigeons for loading into a W29; the pigeons could be vital if the aircraft was forced down and the crew needed to be rescued.

Facing Page: W29 Marine #2287 was the type aircraft for the C2MGHFT version, which carried a wireless transmitter and receiver in addition to the forward-firing gun and flexible gun; it is shown here at Warnemünde. Flight performance was similar to the C3MG version. Marine #2287 was sent to the front the first half of July 1918.

Brandenburg W29 Production Orders					
Order Date	Marine Nos.	Qty	Category	Engine	Notes
17 Jan. 1918	2204	1	C3MG	150 hp Benz Bz.III	Type aircraft
17 Jan. 1918	2205	1	C2MG	185 hp BMW.IIIa	
17 Jan. 1918	2206	1	C3MG	160 hp Mercedes D.III	
13 Apr. 1918	2287–2300	14	C2MGHFT	150 hp Benz Bz.III	2287 was type aircraft
13 Apr. 1918	2501–2506	6	C2MGHFT	150 hp Benz Bz.III	
13 Apr. 1918	2507–2536	30	C3MG	150 hp Benz Bz.III	2715 destroyed before accept.
25 May 1918	2564–2583	20	C2MGHFT	150 hp Benz Bz.III	
30 May 1918	2584–2587	4	C3MG	185 hp Benz Bz.IIIa	
30 May 1918	2588–2589	2	C2MGHFT	185 hp Benz Bz.IIIa	
9 July 1918	2593–2642	50	C3MG	150 hp Benz Bz.III	2625 was C2MGHFT
9 July 1918	2643–2652	10	C3MG	185 hp BMW.IIIa	
9 July 1918	2653–2682	30	C3MG	185 hp Benz Bz.IIIa	
10 Sep. 1918	2730–2759	30	C3MG	185 hp Benz Bz.IIIa	
10 Sep. 1918	2670–2789	30	C3MG	170 hp Mercedes D.IIIa	None completed due to armistice

Above & Below: The W29 prototype before its Marine Number of 2204 was added. The rudder was cut away at the top for clearance for elevator movement and there are no horn balances on the ailerons.

Below: The W29 prototype before its Marine Number was added. This front view shows its clear lines for a floatplane.

Above, Below, & Bottom: The W29 prototype after its Marine Number was added. The rudder has been enlarged and horn balances have been added to the ailerons as a result of flight testing.

Above, Below, & Bottom: W29 Marine #2292 was from the first production batch. The first production batch were category C2MGHFT and the mount for the wind-driven dynamo is seen on the side of the observer's cockpit.

Above: Based so close to the front lines, the Brandenburgs normally took off and landed in formation to provide mutual protection against attack by British fighters. Here *Oblt.z.S.* Christensen takes off in his Marine #2512 in the background.

This W29 was the personal aircraft of *Oblt.z.S.* Friedrich Christiansen. The black letter 'C' for 'Christiansen' in a black diamond painted over a white stripe on the rear fuselage was Christiansen's personal insignia. The aircraft wears standard late-war naval camouflage.

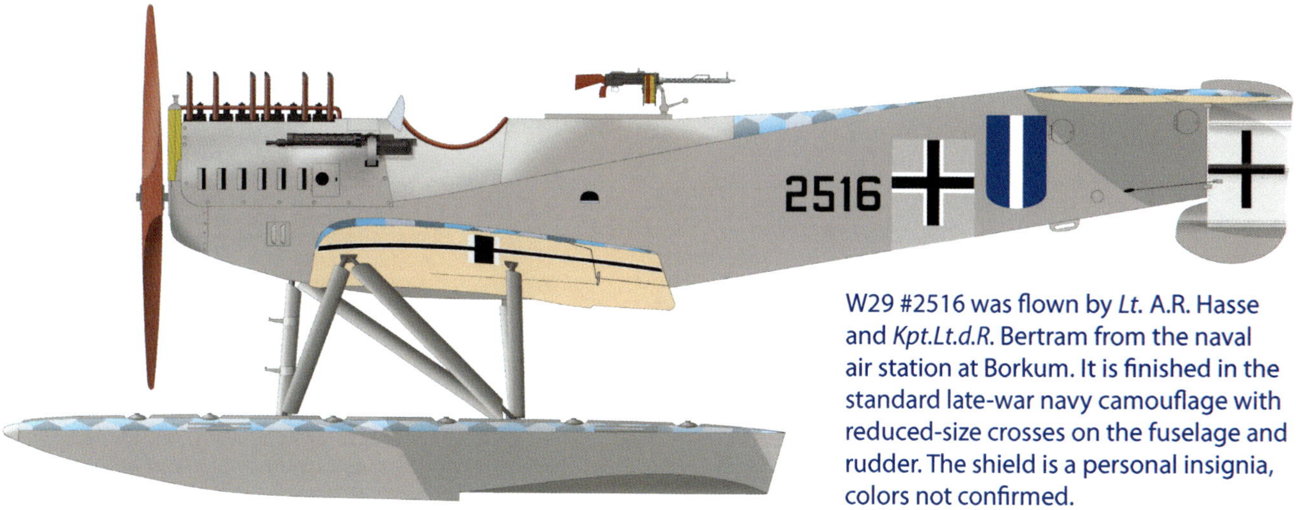

W29 #2516 was flown by *Lt.* A.R. Hasse and *Kpt.Lt.d.R.* Bertram from the naval air station at Borkum. It is finished in the standard late-war navy camouflage with reduced-size crosses on the fuselage and rudder. The shield is a personal insignia, colors not confirmed.

Above: W29, possibly Marine #2655, was fitted with a 245 hp Maybach Mb.IVa. The aircraft was easier to fly but not as maneuverable. A tachometer was in front of the pilot.

Above: Mechanics work on the fixed pilot's guns on this W29.

Above & Below: W29 Marine Number 2670 under evaluation in the UK post-war. British roundels have been applied over the German insignia on the wings.

Above, Below, & Bottom: W29 Marine Number 2670 under evaluation in the UK post-war. The front view above shows the strikingly clean lines of the W29 that gave the type its additional speed compared to the W12.

Above & Top: Two views of the first Austro-Hungarian W29. Built by UFAG, it was designated the W29(U) and assigned the Austro- Hungarian naval serial C1. The pointed radiator was typical for aircraft fitted with a 185 hp Austro-Daimler engine. Fairings on the side of the fuselage enclose the pilot's guns. First flight was on 25 Oct. 1918, too late to see combat.

Right: A flight of Danish license-built W29s, called H.M.Is, over Copenhagen in May 1915. The W29 had a long and successful post-war career in Denmark and Norway.

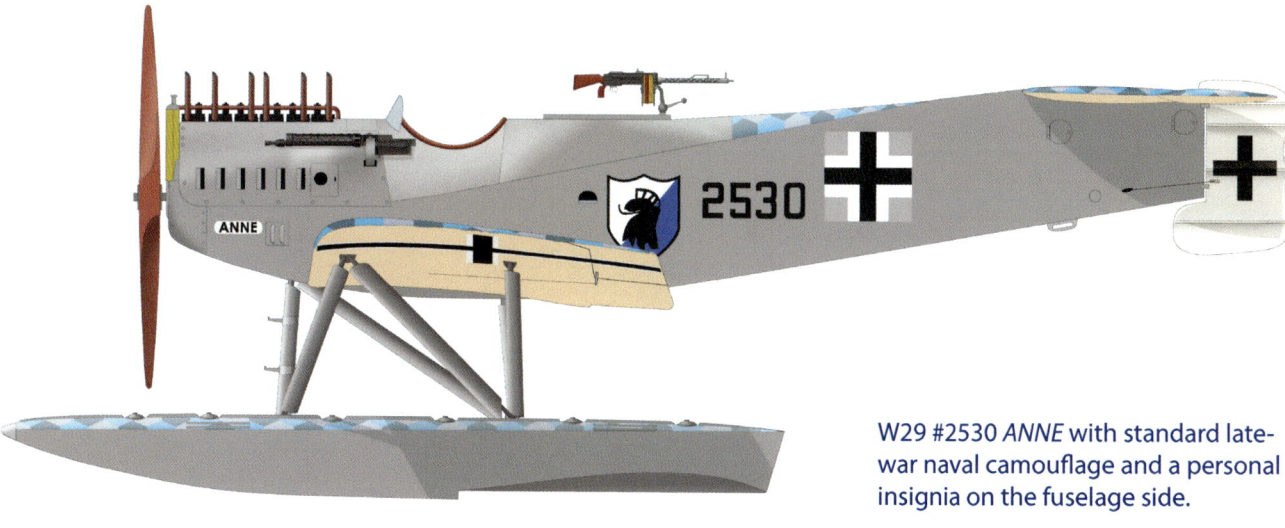

W29 #2530 *ANNE* with standard late-war naval camouflage and a personal insignia on the fuselage side.

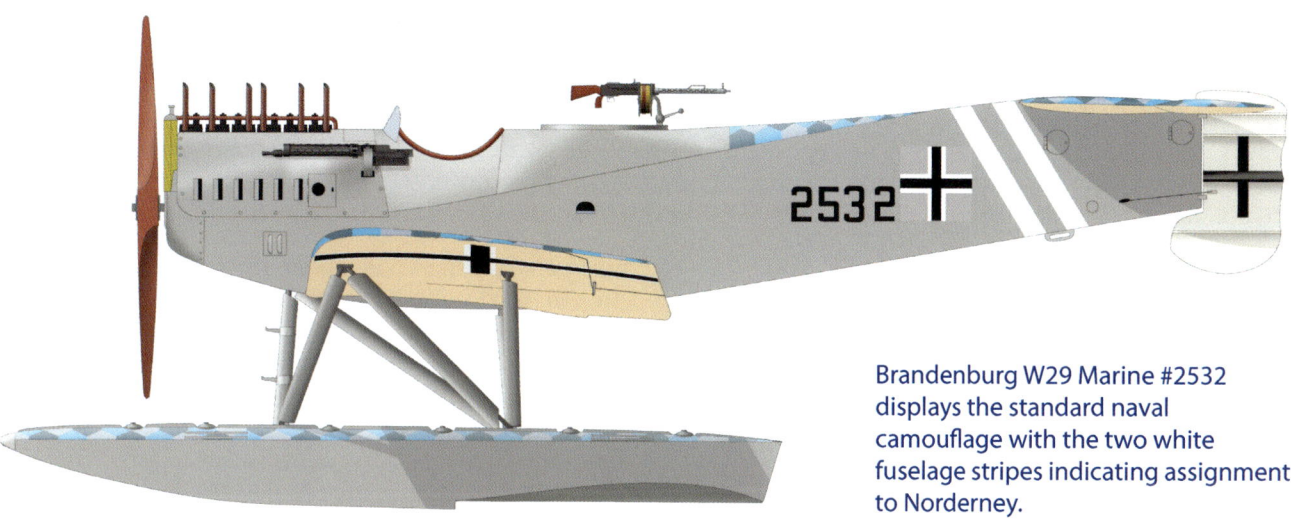

Brandenburg W29 Marine #2532 displays the standard naval camouflage with the two white fuselage stripes indicating assignment to Norderney.

C1 was the first Austro-Hungarian W29(U). Only the rudder, in the national colors with crown, added color to the camouflage.

Brandenburg W33, W34, & W37

Above: The W33 was an enlarged development of the W29 powered by the 260 hp Mercedes D.IVa or similar engines and designed for greater range and endurance. Photographed at Warnemünde on 30 August 1918, these views of W33 Marine #2538 shows the very clean lines of the W33 despite its size. It was finished in the standard late-war naval camouflage.

Just as the W19 was a larger, more powerful development of the W12 biplane for greater range and endurance, the Brandenburg W33 was a larger, more powerful development of the W29 monoplane. The larger size of the W33 enabled it to carry more fuel for a longer range and greater endurance, and its greater power made it as fast. The first three W33s were powered by the 260 hp Maybach Mb.IVa, but the 260 hp Mercedes D.IVa and 275–300 hp Basse & Selve BuS.IVa were also used.

The Brandenburg W29 went into operational service in April/May 1918 and was immediately successful in air combat over the North Sea. The larger, longer-range W33 followed in late summer.

Although only seven W33 aircraft were delivered during the war, they were ordered in three classes; the C3MG, armed with three machine guns, the C2MG HFT with a fixed gun, a flexible gun, and wireless (radio) equipment, and a CK class with two fixed machine guns and a flexible Becker 20mm cannon instead of a Parabellum machine gun for the observer.

Like the earlier W12, W19, and W29, the W33 aircraft were used primarily from North Sea air stations where they performed armed reconnaissance missions and opposed the British flying boats that were performing anti-submarine patrols. The W33 was ordered in very limited

numbers and may have been an interim step to the enlarged W34. Three W34 floatplanes in the C3MG class were also ordered and delivered but were too late to serve operationally before the armistice. Drawings show the W34 had an enlarged gun ring capable of mounting a 20mm Becker cannon. Work on the similar but even larger W37 was halted post armistice and these three aircraft were not completed. However, eleven W37 aircraft, powered by the 260 hp Maybach Mb.IV, were built in Germany post-war and shipped to Sweden for assembly as the Caspar S.I, and others were built in Sweden by the Heinkel company.

Along with the smaller W29, the W33 enjoyed a long and successful postwar career in Norway and Finland. The W33 was produced under license in Finland from 1922–1925, 120 aircraft being built. The 220 hp Benz Bz.IV was not available so 300 hp Fiat A-12bis engines were purchased from France. The first 57 aircraft built had provision for a fixed Vickers gun for the pilot, but these were not fitted to aircraft in service. The observer had a pair of Lewis guns on a flexible gun ring. Known as the IVL A.22 in Finnish service, the type served until 1936. License production of the W33 was also undertaken in Norway, 24 aircraft being delivered during 1920–1929 and serving until 1935. Most Norwegian W33s used the 260 hp Mercedes D.IVa engine.

Above & Top: The first W33 displays its clean, robust lines and great resemblence to the smaller W29.

Brandenburg W33, W34, & W37 Production Orders					
Order Date	**Type**	**Marine Nos.**	**Qty**	**Category**	**Engine**
24 Apr. 1918	W33	2538–2540	3	C3MG	260 hp Mercedes D.IVa
24 Apr. 1918	W33	2541	1	C2MG HFT	260 hp Maybach Mb.IVa
24 Apr. 1918	W33	2542	1	C3MG	275–300 hp BuS.IVa
24 Apr. 1918	W33	2543	1	CK	275–300 hp BuS.IVa
24 Apr. 1918	W33	2726	1	C3MG	260 hp Mercedes D.IVa
29 Aug. 1918	W34	2727–2729	3	C3MG	275–300 hp BuS.IVa
24 Aug. 1918	W37	2723–2725	3	CHFT	220 hp Benz Bz.IV

Left: Finnish W33s were sturdy enough to be operated from ice during winter. These aircraft used the 300 hp Fiat A-12bis, and 120 were built from 1922. The type served until 1936!

Right: Norwegian W33s in operation. Most Norwegian W33s used the 260 hp Mercedes D.IVa engine. Armament was a fixed Vickers for the pilot and a flexible Lewis for the observer. The original nose radiators were replaced with a pair of under-wing Lamblin radiators, altering the aircraft's appearance.

Right: Norwegian W33 F-16 was fitted with a blind flying hood for practice instrument flying. It was delivered to the Norwegian Navy in May 26, 1923 and survived for ten years.

Below & Below Right: Norwegian W33s used Lamblin radiators under the wings instead of the nose radiators of the original design.

Above: The W33 was known as the IVL A.22 in Finnish service, this example survives in the Finnish aviation museum. The wing was complex and presented a manufacturing challenge to the Finns when they undertook license production; a lot of technology transfer took place in 1922 between Germany and Finland.

Below: Marine #2538, the first Brandenburg W33.

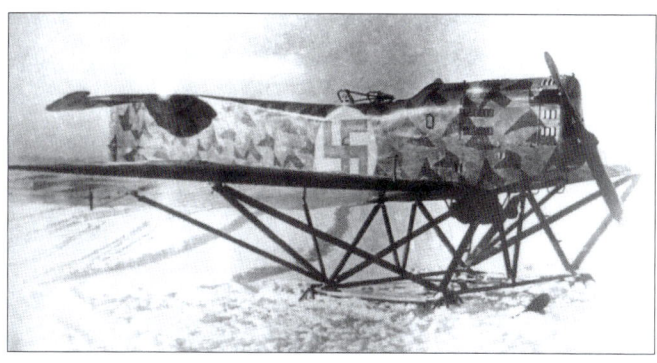

Above: Some Finnish W33/IVL A.22 aircraft were fitted with skis. Armament was a pair of flexible Lewis guns.

Left: Experimental camouflage on Finnish W33/IVL A.22.

Friedrichshafen FF63

Above: Powered by the 200 hp Benz Bz.IV, the Friedrichshafen FF63 appears to be a fair design that just does not match the innovative structural design of the W29 and W33 that were its competitors. The additional bracing struts above the wing and the struts above the fuselage added weight and drag that the Brandenburg monoplanes avoided. The tail gave the gunner a good field of fire, but again the Brandenburg designs had a better field of file.

Friedrichshafen, who manufactured more seaplanes than any other German company, produced a competitor to the Brandenburg monoplanes, the FF63. The engine was the 200 hp Benz Bz.IV; virtually no other technical data on the type survives, and even its Marine Number is not known.

The FF63 was not as innovative in its structural design as the Brandenburg monoplanes and only a single FF63 was built. The radiator above the engine was surely fitted to expedite early flight testing; this installation produced far too much drag for a production aircraft and also obstructed vision.

Below: Only one Friedrichshafen FF63 was built. The radiator mounted above the engine was surely to expedite flight testing and would certainly have been replaced by an installation of lower drag had it gone into production.

Junkers CLS.I

Above: The all-metal Junkers CLS.I was a floatplane development of the CL.I two-seat fighter. The CL.I was produced for the army and used postwar. The CLS.I was powered by a 200 hp Benz Bz.IV engine. The CLS.I had a top speed of 180 km/h, which was somewhat faster than the W29. The corrugated metal skin was characteristic of early Junkers designs.

A more serious competitor was the all-metal Junkers with factory designation J11, military designation CLS.I, a floatplane development of the Junkers CL.I two-seat fighter. A fixed fin was added to compensate for the side area of the floats in front of the center of gravity, but insufficient stability required modifications that continued even after the war, when the type was modified for civil use. The all-metal structure resisted the maritime environment better than did wood structures like those of the Brandenburgs, but the type was too late for wartime service.

Right: Marine #7501 was a floatplane adaption of the all-metal Junkers CL.I two-seat fighter. The CL.I landplane had no fixed fin, but one was required on the CLS.I due to the destabilizing effect of the floats. The CLS.I tail surfaces underwent extensive modifications to achieve good flying characteristics, but the type arrived too late for wartime service.

Zeppelin-Lindau (Dornier) Cs.I

Above: The Cs.I on a compass platform. The bracing wires and box radiators on the fuselage side produced too much drag for the design to fulfill its designed performance despite its 195 hp Benz V-8 engine.

Another serious competitor for two-seat floatplane fighter production was the Zeppelin-Lindau (Dornier) Cs.I monoplane powered by a 195 hp Benz V-8. This aircraft was also all metal except for fabric-covered wing and tail surfaces. The drag-producing box radiators mounted on the fuselage sides for testing undoubtedly resulted in lower speed than the design was potentially capable of, but were likely an expedient to speed up flight testing. The single Cs.I flown, Marine #8502, had not completed testing when the war ended.

Below: Dornier Cs.I Marine Numbe 8502 was the only one of the three airframes to fly.

Above: The Zeppelin-Lindau (Dornier) CS.I is seen during testing with the spinner removed. It still has the drag-producing ear radiators.

Below: The CS.I was intended to use a nose radiator as seen here. This installation produced less drag than the ear radiators and enabled the CS.I to attain greater speed.

In Retrospect

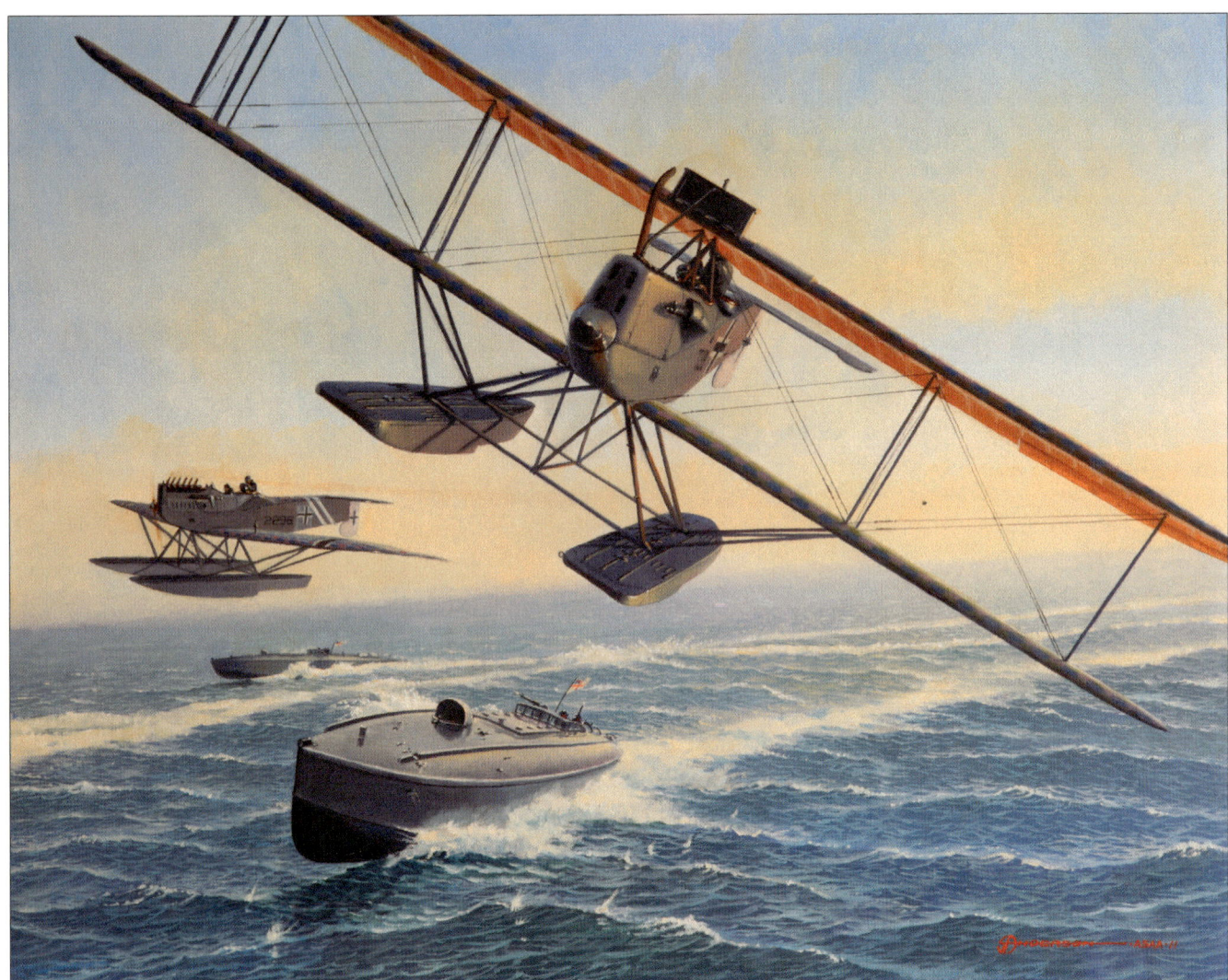

Above: *Race for Life.* This striking painting by Steve Anderson depicts the action on August 11, 1918 when Brandenburg fighters attacked six British Motor Torpedo Boats (MTBs) operating against German minesweepers. Three of the MTBs were sunk and the remaining three were damaged so badly they beached themselves in neutral Holland to avoid sinking. Brandenburg W19 Marine #2249 shown here was one of the fighters that sank an MTB.

As the course of the maritime air war changed, so did the needs and procurement policies of the German Navy. Evolution of naval combat aircraft paralleled that of army aircraft to a large extent. Reconnaissance was the primary role of maritime aviation throughout the war, just as it was for the army. Initially the primitive reconnaissance seaplanes were unarmed. Airplanes of all types were evolving into larger, more powerful craft of longer range, greater speed, and the ability to carry increased loads, including armament. As air combat evolved from an oddity to routine, the reconnaissance seaplanes were armed.

Soon the Navy recognized the need for fighters to defend naval air stations against reconnaissance and bombing. Single-seat floatplane fighter prototypes were built, and all three types that entered production were modifications of Army aircraft already in use. These fighters were able to engage hostile seaplanes and two-seaters on favorable terms, but were inferior in performance to contemporary landplanes due to the weight and drag of their floats. Their range and endurance were too short for offensive operations, and over-water navigation in a single-seat aircraft was also problematic.

These realities and the desire to conduct longer-

Above: As illustrated here, longer-range W19s would frequently patrol with W12s and W29s, which sometimes waited on the surface until the W19 found a target and returned to lead the flight to it.

range, offensive operations led the Navy to procure two-seat floatplane fighters. In addition to their longer range and endurance, the second crewman was an invaluable help with over-water navigation, and his flexible machine gun improved combat effectiveness. In addition, some of these aircraft carried a wireless transmitter/receiver operated by the observer, something not possible for a single-seat aircraft.

Brandenburg's innovative W12 performed better than its competitors and was produced in significant numbers for a naval aircraft. The great operational success of the two-seat W12 over the North Sea led to all subsequent production Brandenburg floatplane fighters being of similar configuration. Competing manufacturers were unable to design comparable aircraft until the innovative, all-metal Junkers CLS.I and Dornier Cs.I designs, which were too late for combat.

The North Sea was not only a seaward continuation of the trench lines on the Western Front, German submarines also had to transit through it to their patrol areas, attracting British airships and flying boats on anti-submarine operations. With the appearance of the Brandenburg W12 the Germans were now able to effectively counter those operations.

For even greater range and endurance than the W12, a larger, more powerful aircraft was needed, and Brandenburg's W19, basically an enlarged W12, was produced in response to that need.

Although the W12 was faster than the big Felixstowe flying boats that were its frequent antagonists, its closing speed was not as great as desired. A fleeing Felixstowe could sometimes evade a W12 long enough that the W12 might have to turn back due to low fuel state before completing the interception. So, for greater speed, the biplane W12 was developed into the monoplane W29. The reduced drag made the W29 faster, which enabled it to intercept the Felixstowes much more quickly to ensure they did not escape.

Just as there was a need for a larger, longer-range development of the W12, the same was true of the W29, and the W33 combined the W29 monoplane's configuration with the larger size, greater fuel capacity, and increased range and endurance of the W19. The W29 and W33 were difficult to distinguish during combat, and there were few of them, resulting in the first type to arrive at the front, the W29, getting all the recognition.

The W33 was to have been followed by the further enlarged W34, but the W34 did not enter service before the armistice. Postwar the similar W37 design was built in Germany and assembled in Sweden as the Caspar S.1. Of 305 Brandenburg monoplanes ordered, 209 were delivered during the war.

The W29 and W33 monoplane floatplane fighters were robust, maneuverable, and fast for their time, and served successfully for many years after the war in demanding Nordic environments, being built under license by Denmark, Norway, and Finnland.

An interesting sidelight to the story of German seaplane fighters is the great variety of types produced considering their comparatively limited production quantities.

Above: British non-rigid airship C27 goes down in flames after an attack by Brandenburg W12s led by *Oblt.z.S.* Christiansen over the North Sea on 11 December, 1917.

Left: A W29 attacks British Submarine *C25* which was caught on the surface on 6 July 1918. Armed only with machine guns, the Brandenburgs penetrated the pressure hull, preventing the submarine from safely submerging. Their gunfire killed the captain and five crewmen and damaged the submarine so badly it was forced to return to port for repairs. For this action *Oblt.z.S.* Friedrich Christainsen was credited for a victory over *C25*.

Left: British submarine *C25* under attack by Brandenburg W29s on 6 July 1918.

Above & Below: Two formations of Brandenburgs hunt over the North Sea. Brandenburgs often patrolled in formations of five aircraft; the trailing aircraft in the formation in the lower photo took the photo of the others.

Above: A pair of W29s in flight display their two identification stripes on their rear fuselages, which indicate assignment to *Seeflugstationen* Norderney.

Above: A W29 of Zeebrugge's 1-*C.Staffel* strafes Felixstowe No.4305 after it was forced down on July 31, 1918.

Above: Felixstowe No.4305 burns. The Brandenburgs earned their reputation as "The Hornets of Zeebrugge."

Below: Brandenburg W29s operating with German warships in the North Sea.

Above: Brandenburg W29 Marine Number 2536 wears an interesting personal insignia on its aft fuselage.

Below: The Brandenburgs did not have things all their own way; here a W19 burns on the water. It may have been forced down by combat damage or mechanical fault, but because the Marine Number for researching the records is not quite legible, the exact cause is unknown.

Vernichtung eines deutschen Kampfflugzeuges

Afterword

Since the Red Baron did not fly floatplane fighters, we end with the most famous floatplane fighter ace and leader of them all, *Oblt.z.S.* Friedrich Christiansen in a formal *Pour le Mérite* portrait at right, reproduced as Sanke card 609. Christiansen took command of *Seeflugstation Zeebrugge* in September 1917 and is credited with at least 13 victories, including airship *C27*, submarine *C25* (which survived), eight twin-engine flying boats, a Sopwith Pup, and two Short 184 floatplanes. He also set the Dutch schooner *Meeuw* on fire April 21, 1918.

Acknowledgements

I want to especially thank Colin Owers for providing many of the contemporary aircraft photos from WWI; his support has been invaluable.

Cover and chapter paintings © Steve Anderson. To learn more about Steve Anderson's art, including how to purchase, please see: www.anderson-art.com

Color aircraft profiles © Bob Pearson. Purchase Bob's CD of WWI aircraft color profiles for $50 USD/Cdn, 40 Euros or 30 GBP with airmail postage included. Paypal to Bob at bpearson@kaien.net or email him for an address to mail a cheque or IMO.

For our aviation books in print and electronic format, please see our website at: www.aeronautbooks.com.

You may contact me at jherris@verizon.net.

Unser erfolgreicher Seekampfflieger Oberleutnant Christiansen

609
Postkartenvertrieb W.Sanke
BERLIN N 37
Nachdruck wird gerichtlich verfolgt!

Bibliography

Books

Gray, Peter, and Thetford, Owen, *German Aircraft of the First World War*, second revised edition, New York: Doubleday & Company, Inc., 1970.

Franks, Norman L.R., Bailey, Frank W., and Guest, Russell, *Above the Lines*, London, Grub Street, 1993.

Grosz, Peter M., *Albatros W4*, Berkhamsted, Albatros Publications, 1995.

Grosz, Peter M., *Brandenburg W12*, Berkhamsted, Albatros Publications, 1997.

Grosz, Peter M., *Brandenburg W29*, Berkhamsted, Albatros Publications, 1996.

Imrie, Alex, *German Naval Air Service*, London, Arms and Armour Press, 1989.

Nowarra, Heinz J., *Marine Aircraft of the 1914–1918 War*, Letchworth, Harleyford Publications Limited, 1966.

Articles

Alexandrov, Andrei, and Kintner, Ron, "The Battle of Angernsee" *Over the Front* Vol.18 No.4, Winter 2003, p.292–329.

Grosz, Peter M., "German Navy Seaplane Serial Numbers and Classification Codes" *Over the Front* Vol.14 No.1, Spring 1999, p.4–15.

Grosz, Peter M., "Rare Birds: Rumpler 6B1" *Over the Front* Vol.18 No.1, Spring 2003, p.74–85.

Grosz, Peter M., "Rare Birds: Rumpler 6B2" *Over the Front* Vol.18 No.4, Winter 2003, p.340–349.

Herris, Jack, "Rare Birds: Hansa-Brandenburg KDW and Derivatives" *Over the Front* Vol.24 No.3, Autumn 2009, p.196–216.

Herris, Jack, "Rare Birds: Hansa-Brandenburg W.19 and W.33–W37 Monoplanes" *Over the Front* Vol.26 No.4, Winter 2011, p.292–311.

Kintner, Ron, "The Brandenburg KDW Operational History" *Over the Front* Vol.24 No.3, Autumn 2009, p.217–229.

Kintner, Ron, "Hansa-Brandenburg W.19 Operational History" *Over the Front* Vol.26 No.4, Winter 2011, p.312–327.

Miller, Thomas G. (ed.), "The Hornets of Zeebrugge: Annotated Excertps from the War Diary of Seeflugstation Flanders I, 1914–1918, *Cross & Cockade* Vol.11 No.1, Spring 1970.

Owers, Colin, "Rare Birds: The Hansa-Brandenburg Monoplanes" *Over the Front* Vol.21 No.4, Winter 2006, p.330–377.

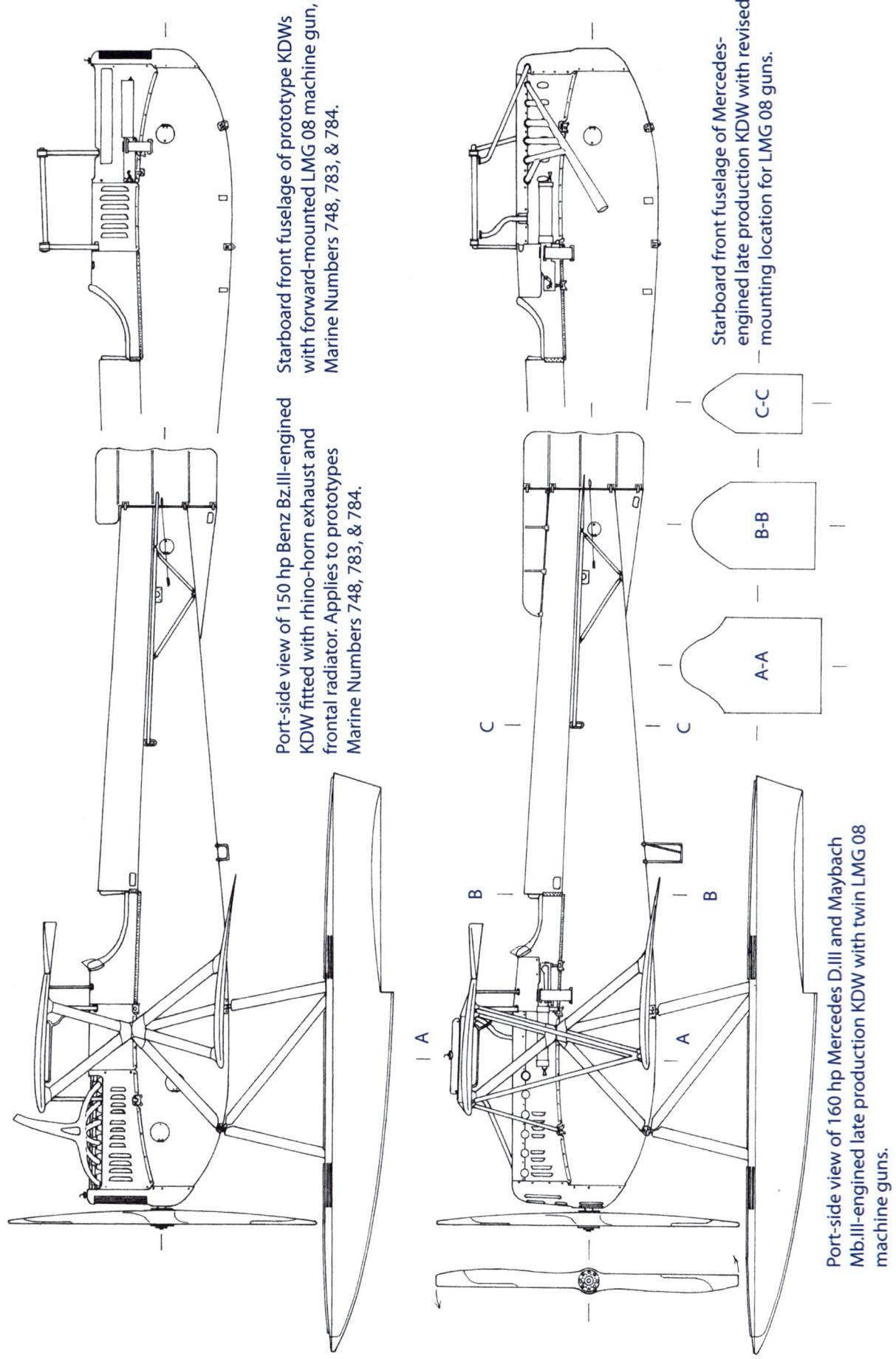

Starboard front fuselage of prototype KDWs with forward-mounted LMG 08 machine gun, Marine Numbers 748, 783, & 784.

Port-side view of 150 hp Benz Bz.III-engined KDW fitted with rhino-horn exhaust and frontal radiator. Applies to prototypes Marine Numbers 748, 783, & 784.

Starboard front fuselage of Mercedes-engined late production KDW with revised mounting location for LMG 08 guns.

Port-side view of 160 hp Mercedes D.III and Maybach Mb.III-engined late production KDW with twin LMG 08 machine guns.

Brandenburg KDW

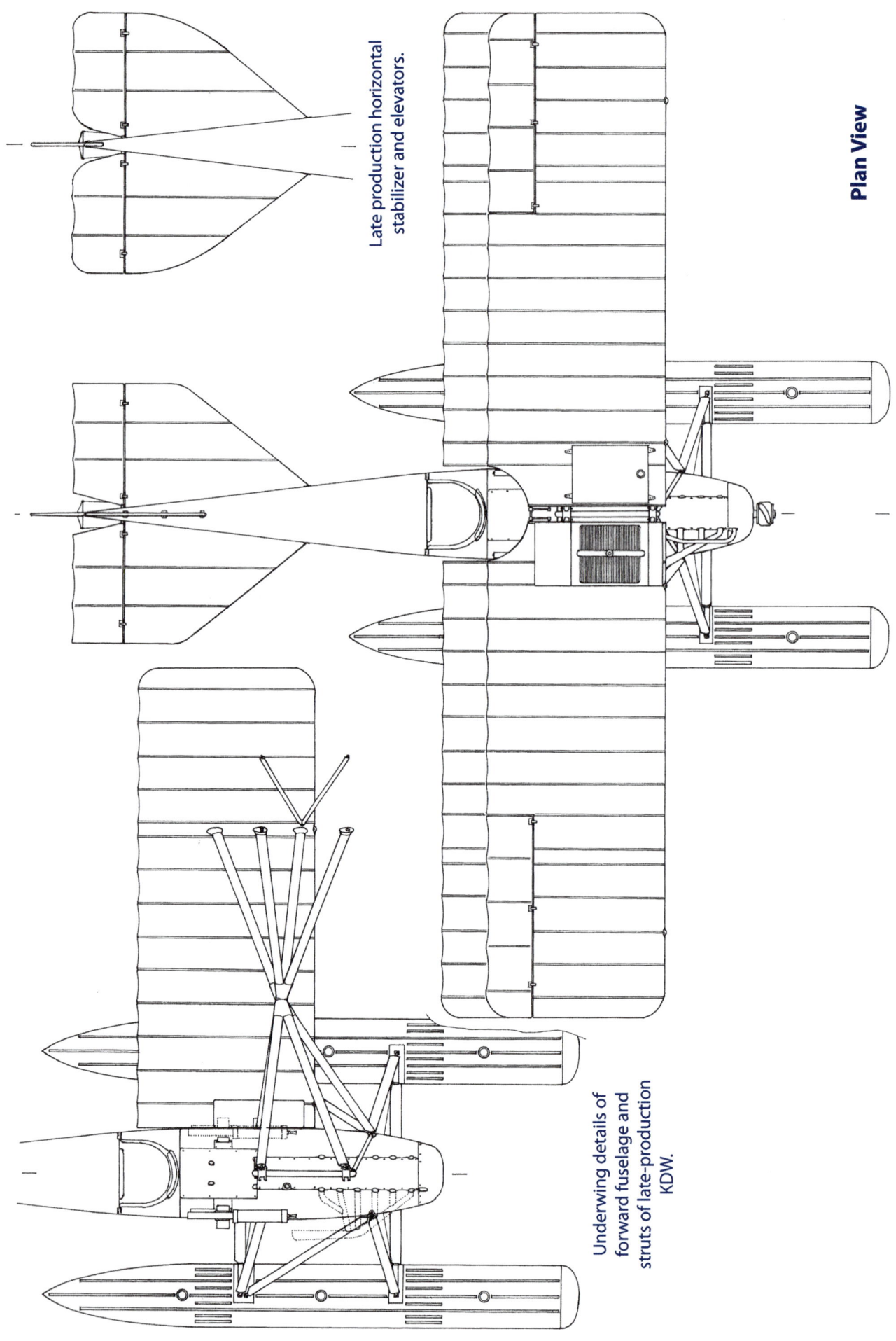

Late production horizontal stabilizer and elevators.

Plan View

Underwing details of forward fuselage and struts of late-production KDW.

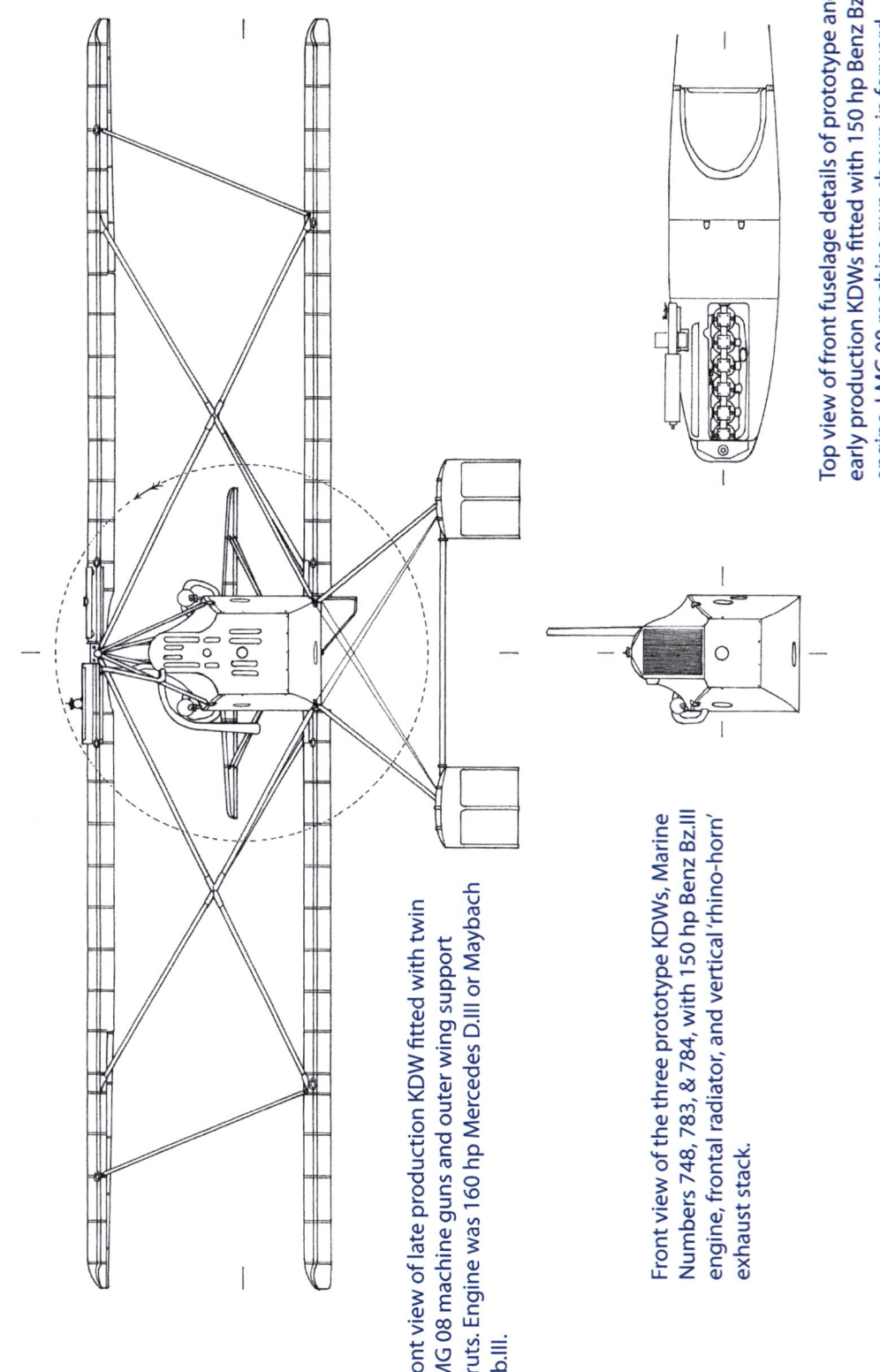

Front view of late production KDW fitted with twin LMG 08 machine guns and outer wing support struts. Engine was 160 hp Mercedes D.III or Maybach Mb.III.

Front view of the three prototype KDWs, Marine Numbers 748, 783, & 784, with 150 hp Benz Bz.III engine, frontal radiator, and vertical 'rhino-horn' exhaust stack.

Top view of front fuselage details of prototype and early production KDWs fitted with 150 hp Benz Bz.III engine. LMG 08 machine gun shown in forward position used by the three prototypes, Marine Numbers 748, 783, & 784.

Brandenburg W16

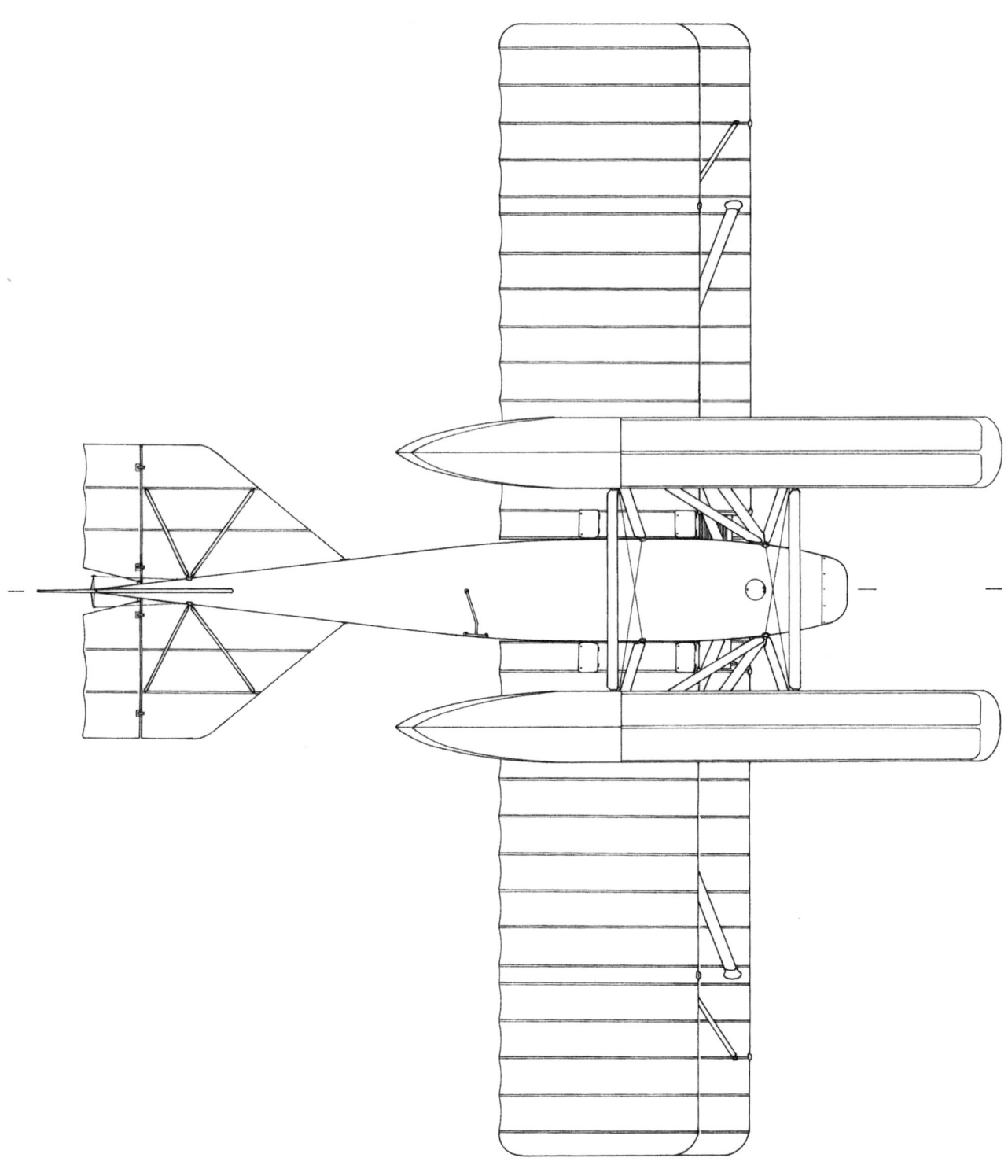

Underside

Brandenburg W16

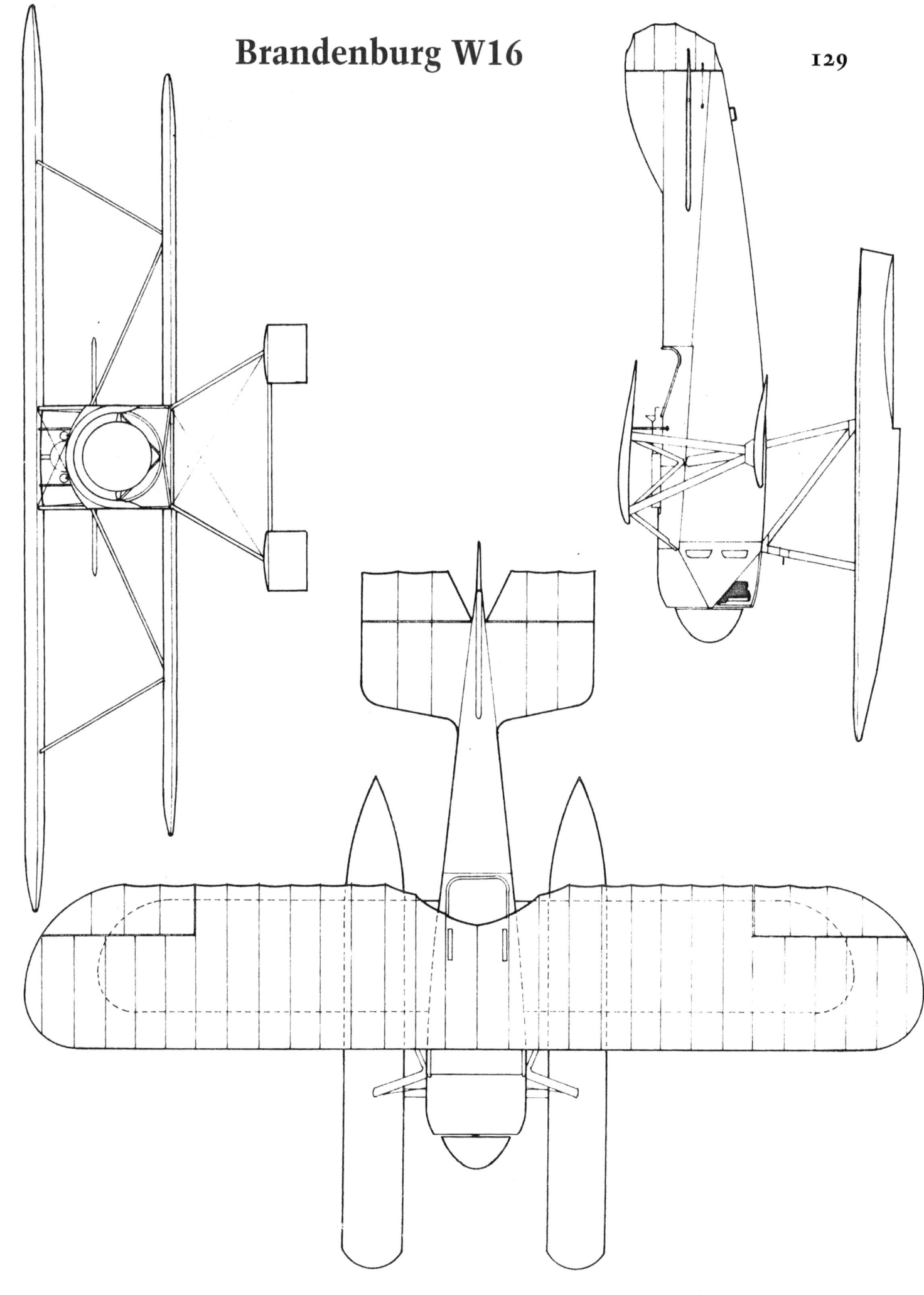

Brandenburg W25

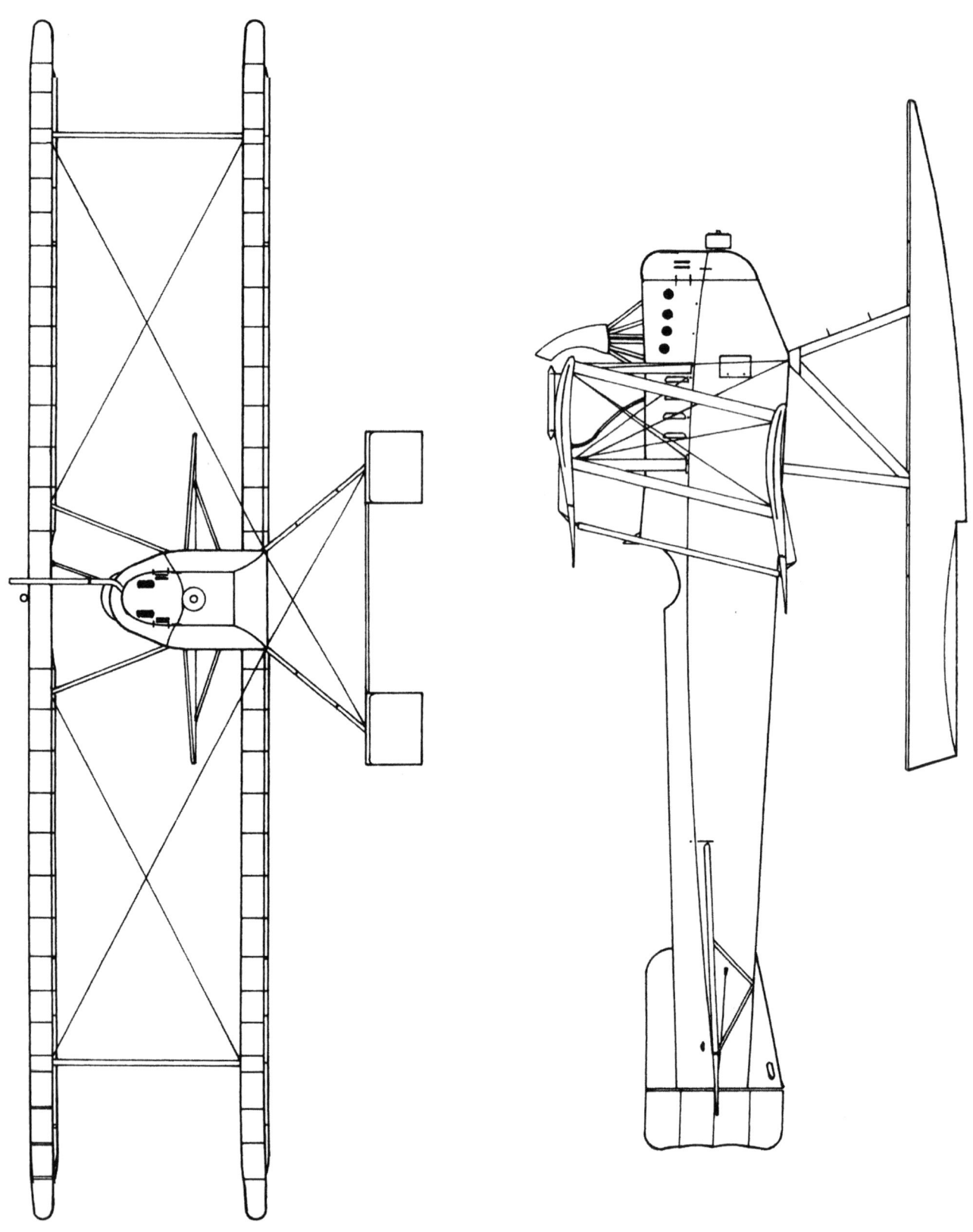

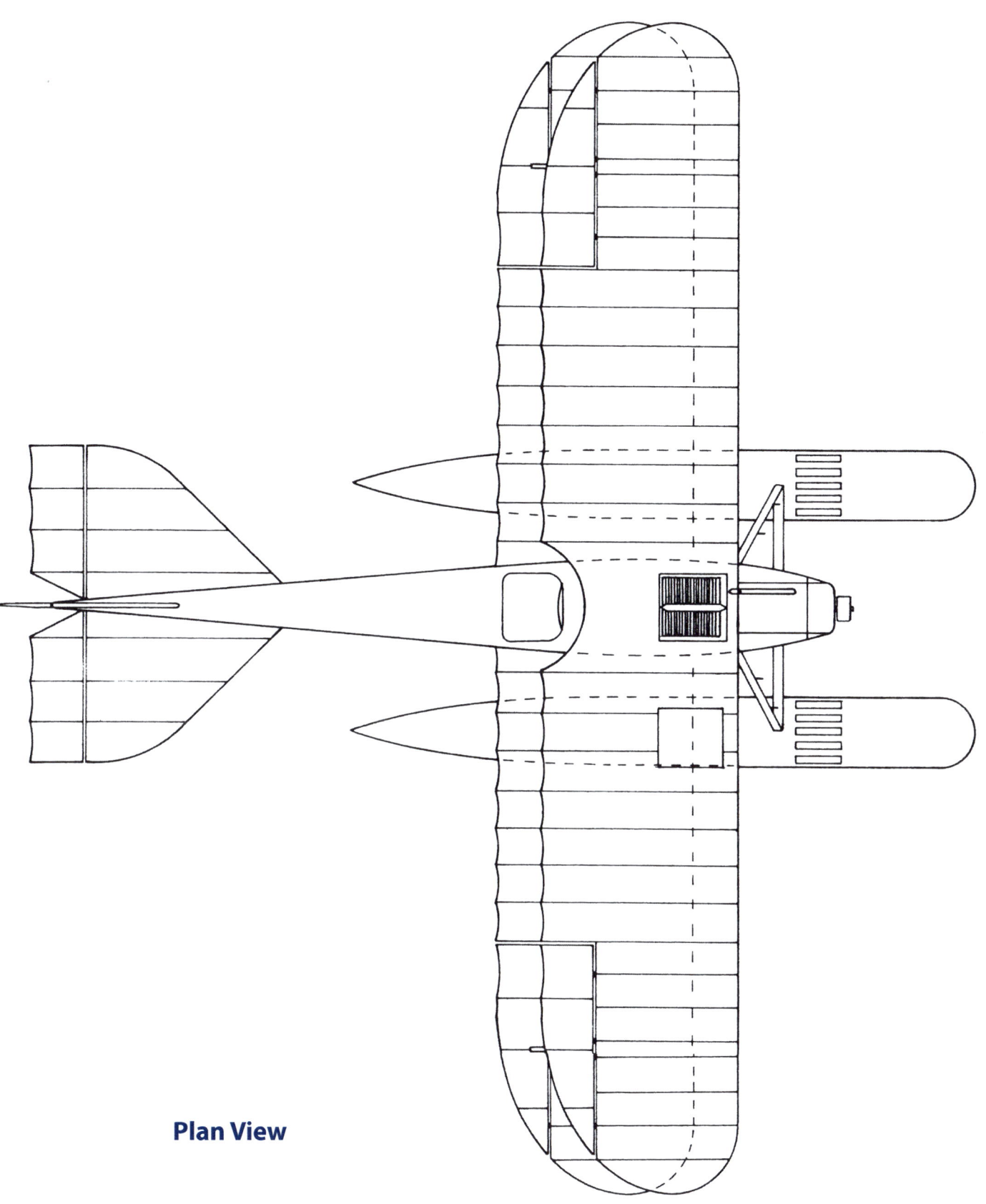

Plan View

Rumpler 6B1

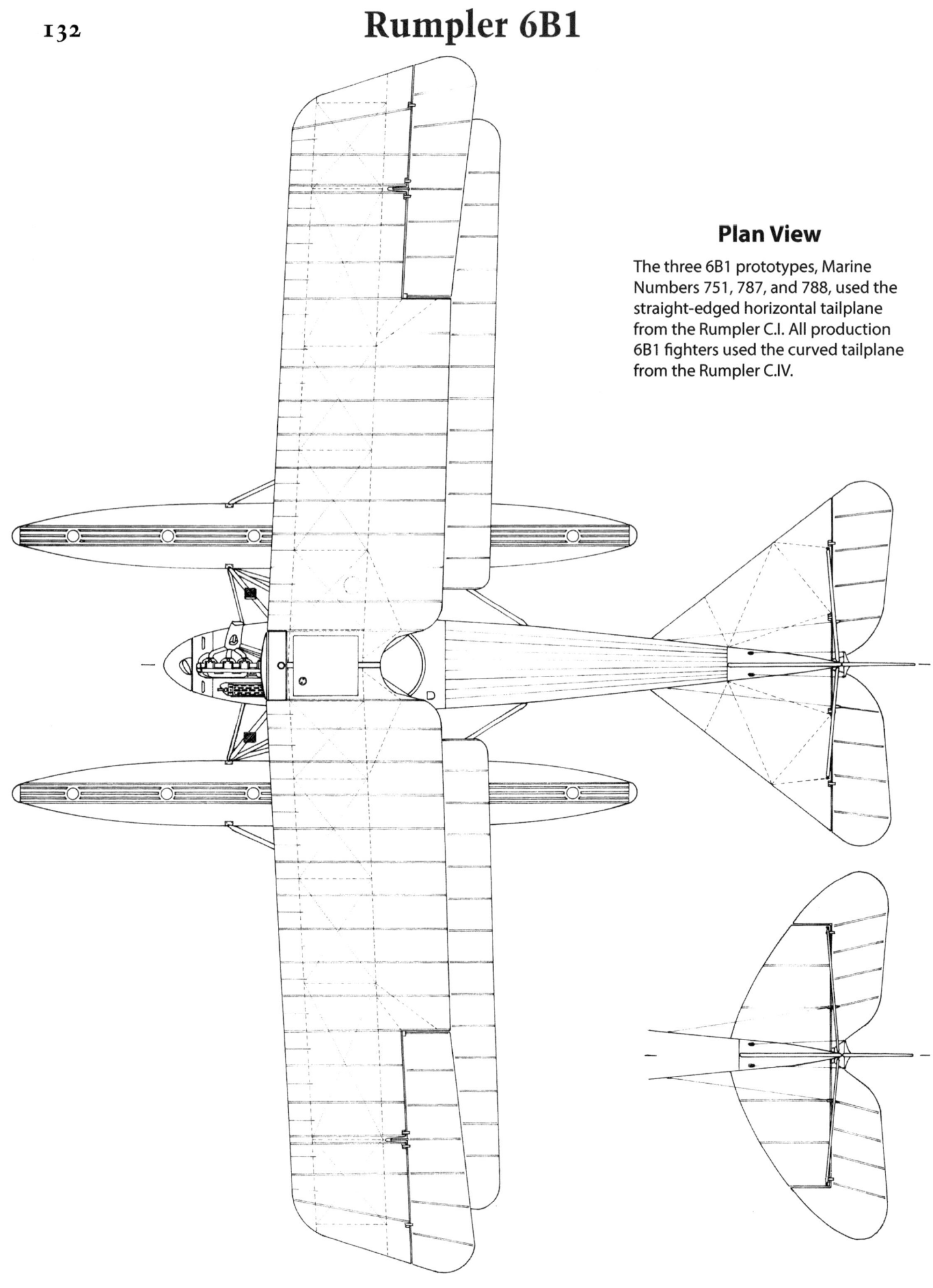

Plan View

The three 6B1 prototypes, Marine Numbers 751, 787, and 788, used the straight-edged horizontal tailplane from the Rumpler C.I. All production 6B1 fighters used the curved tailplane from the Rumpler C.IV.

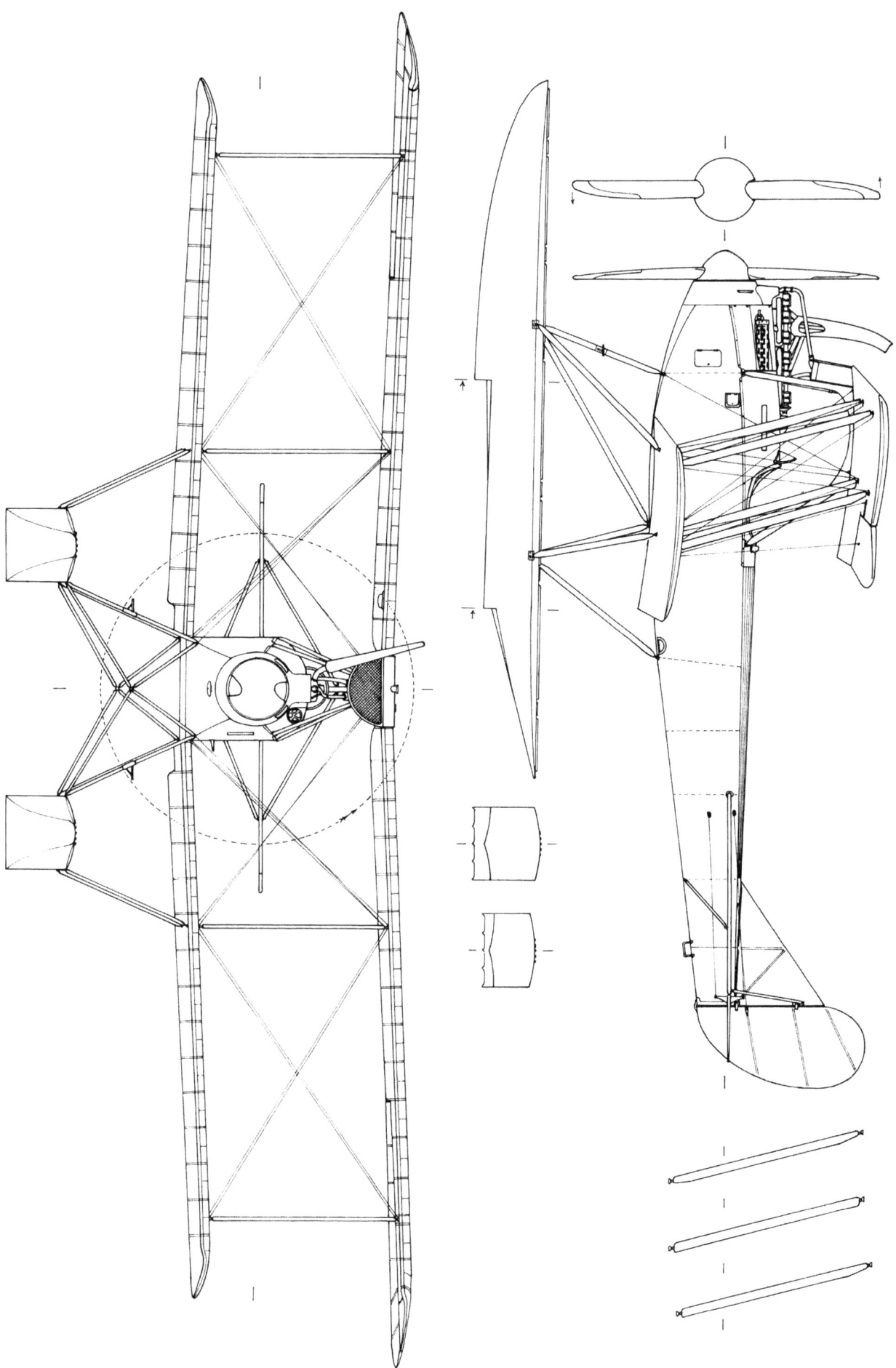

Rumpler 6B1

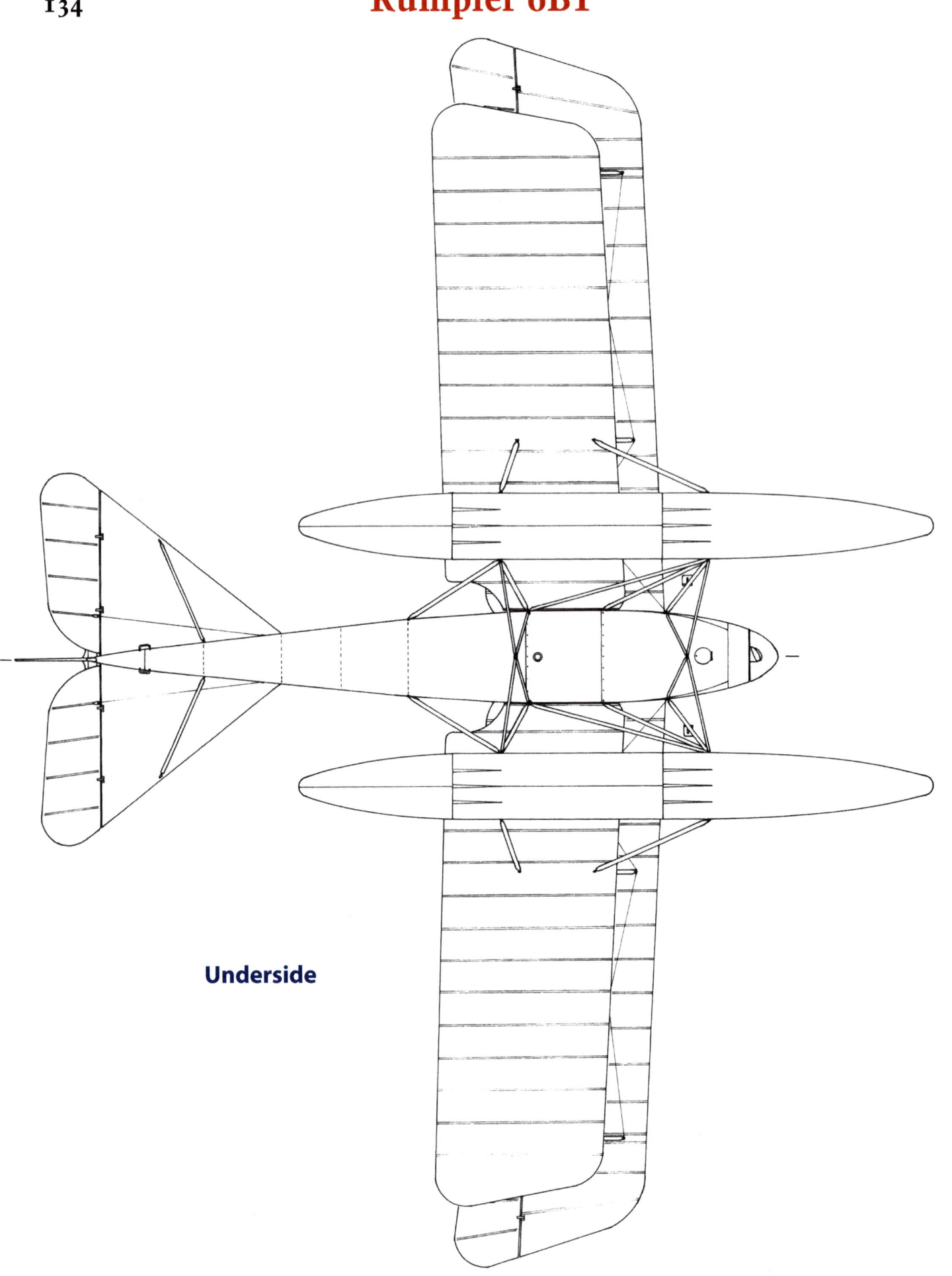

Underside

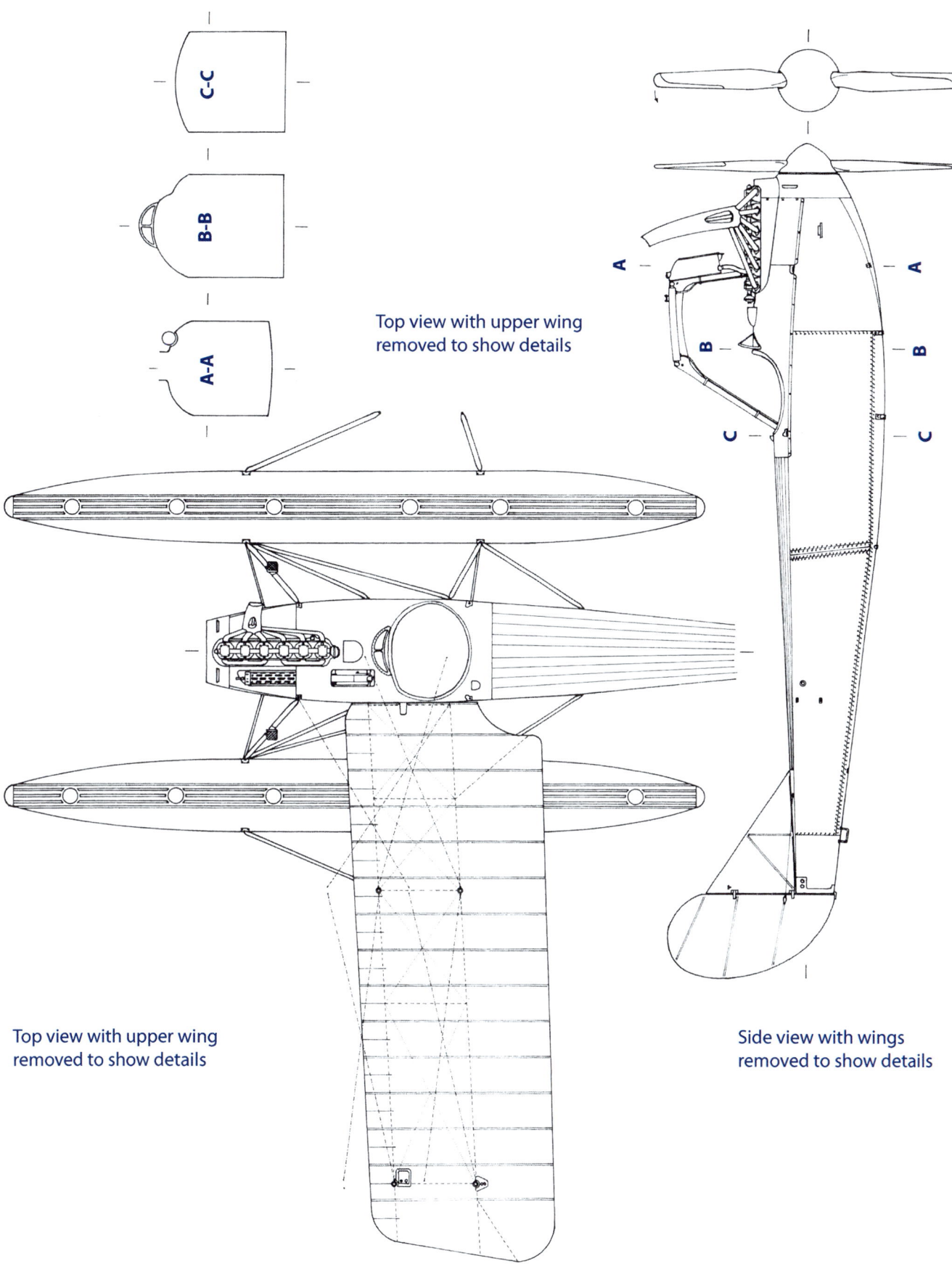

C-C

B-B

A-A

Top view with upper wing
removed to show details

Top view with upper wing
removed to show details

Side view with wings
removed to show details

A

B

C

Side view with wings & tail
removed to show details

Rumpler 6B2

Plan View

Rumpler 6B2

Rumpler 6B2

Underside

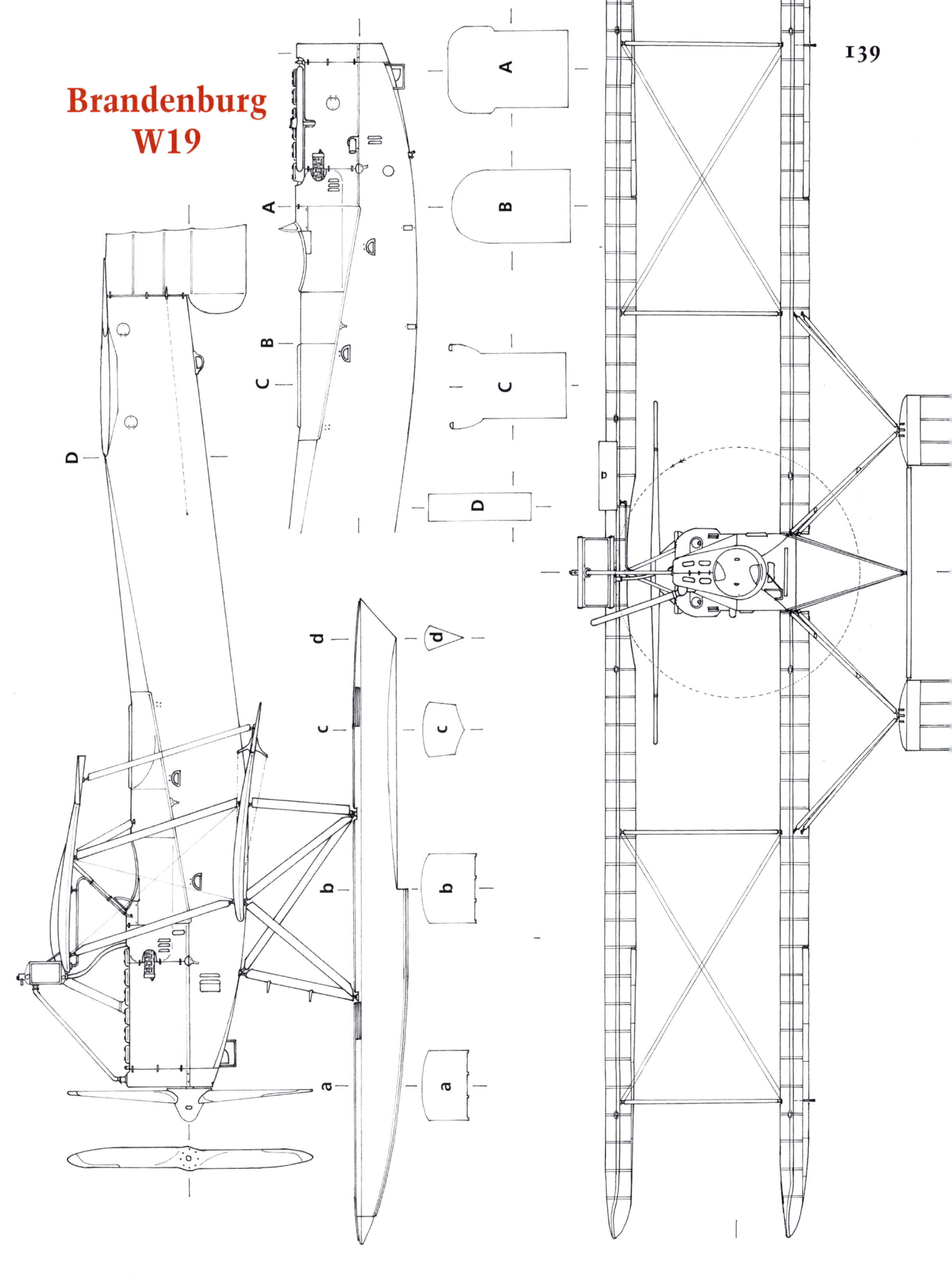

Brandenburg W19

Brandenburg W19

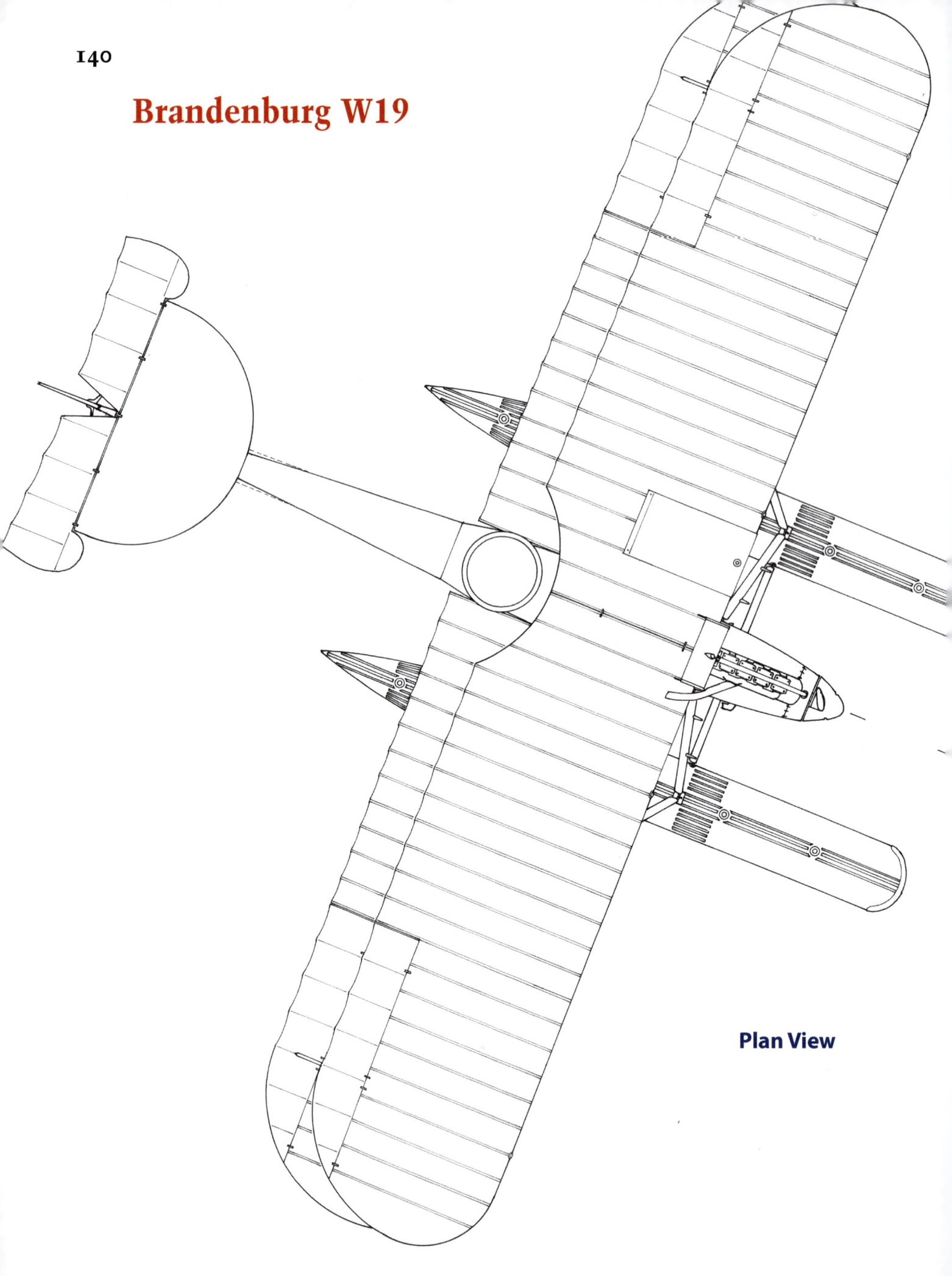

Plan View

Brandenburg W19

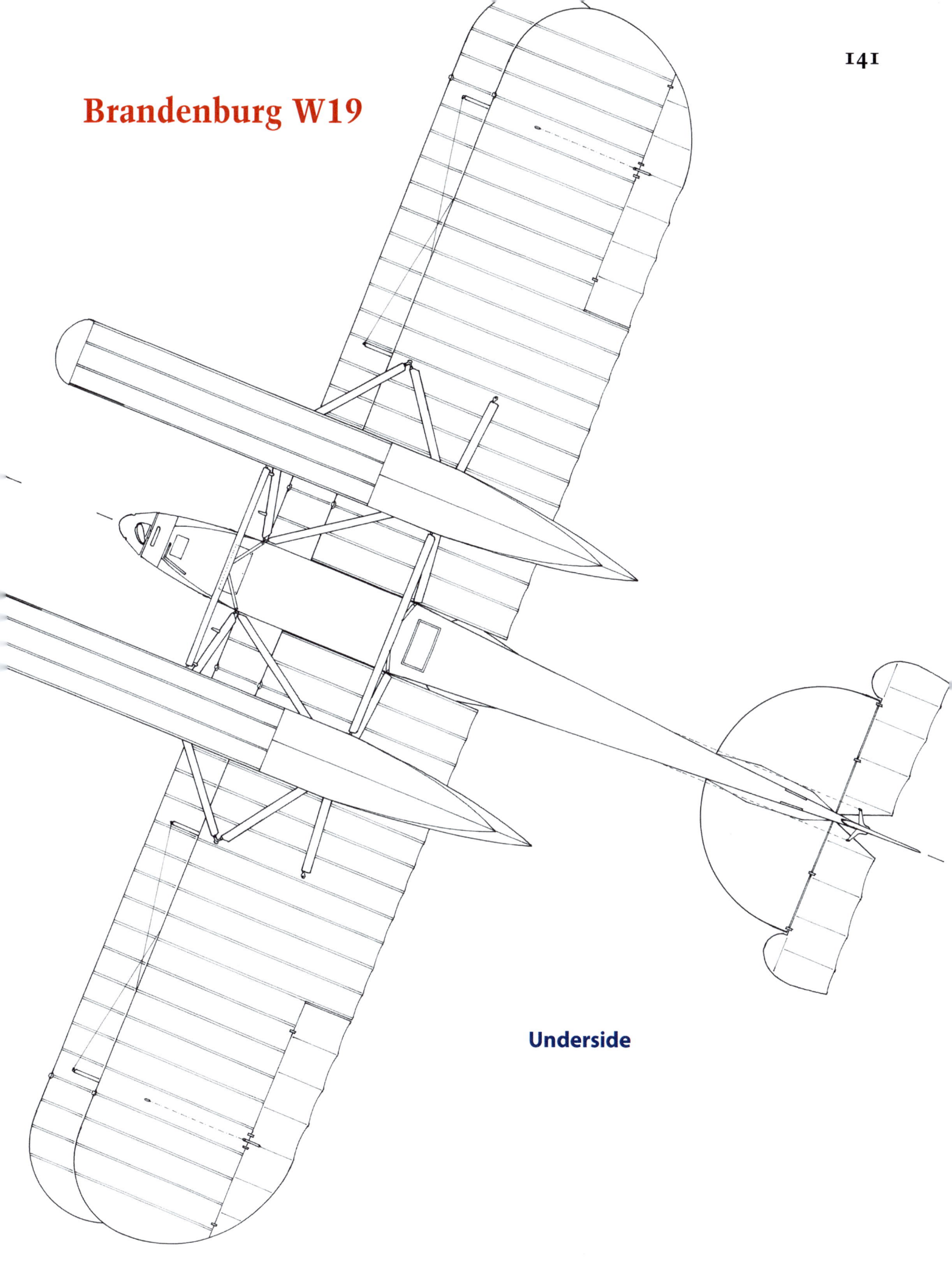

Underside

Brandenburg W19

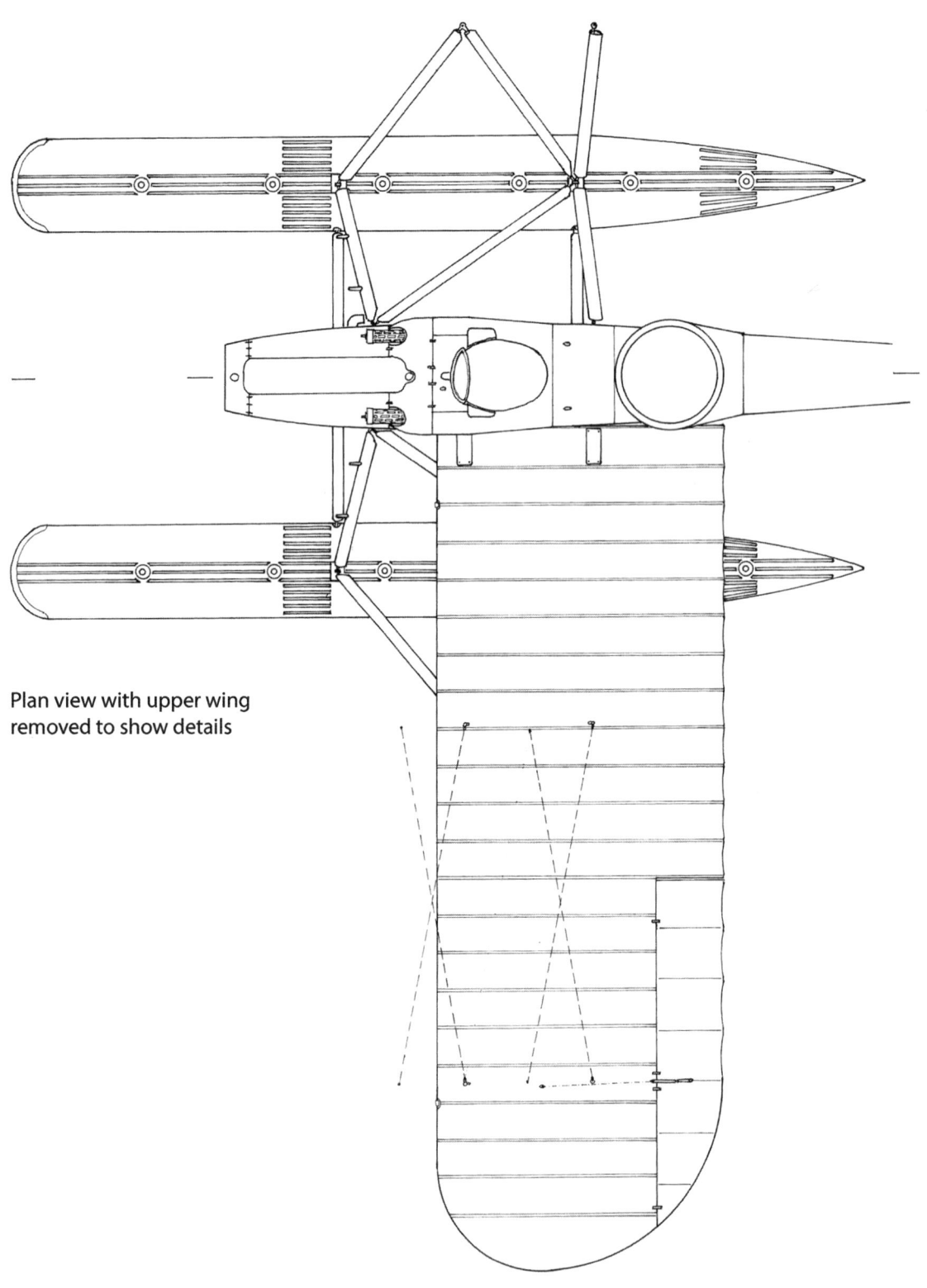

Plan view with upper wing
removed to show details

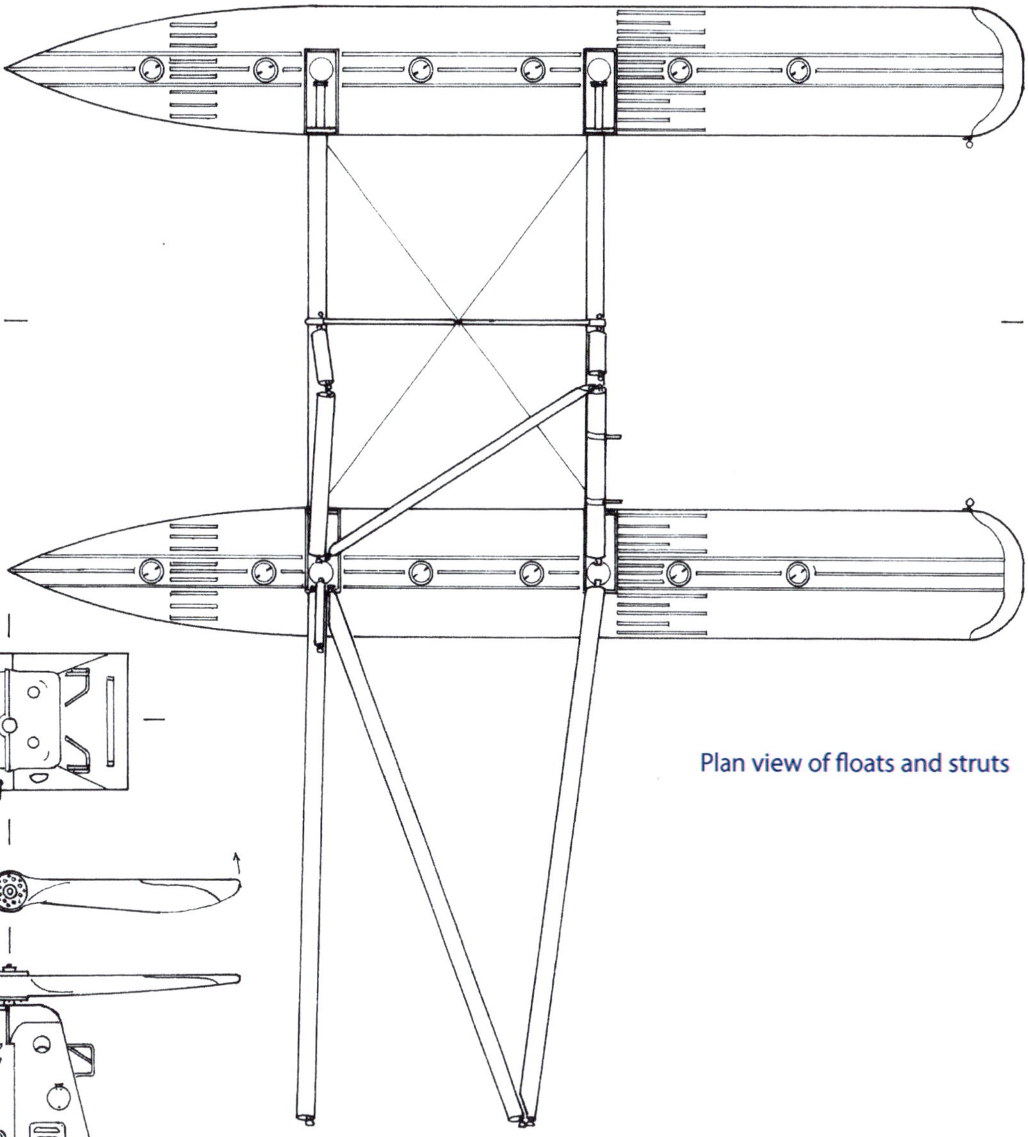

Plan view of floats and struts

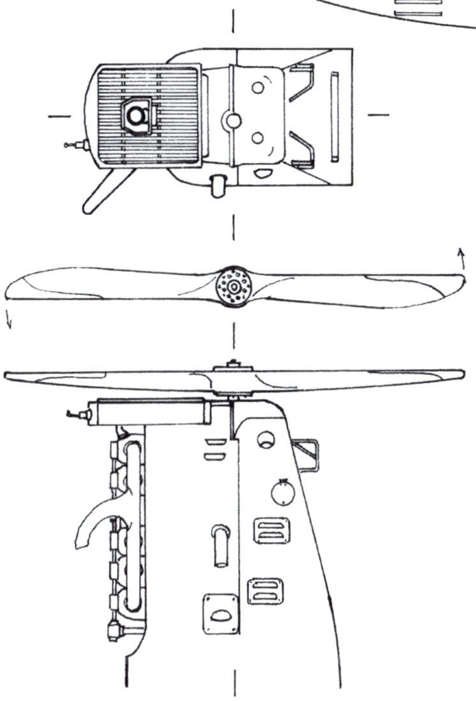

Nose details

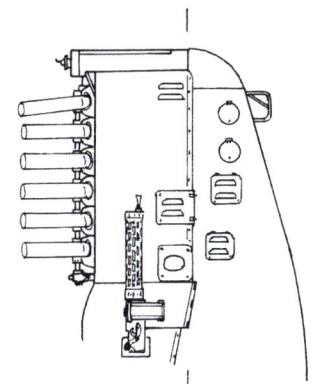

Brandenburg W33

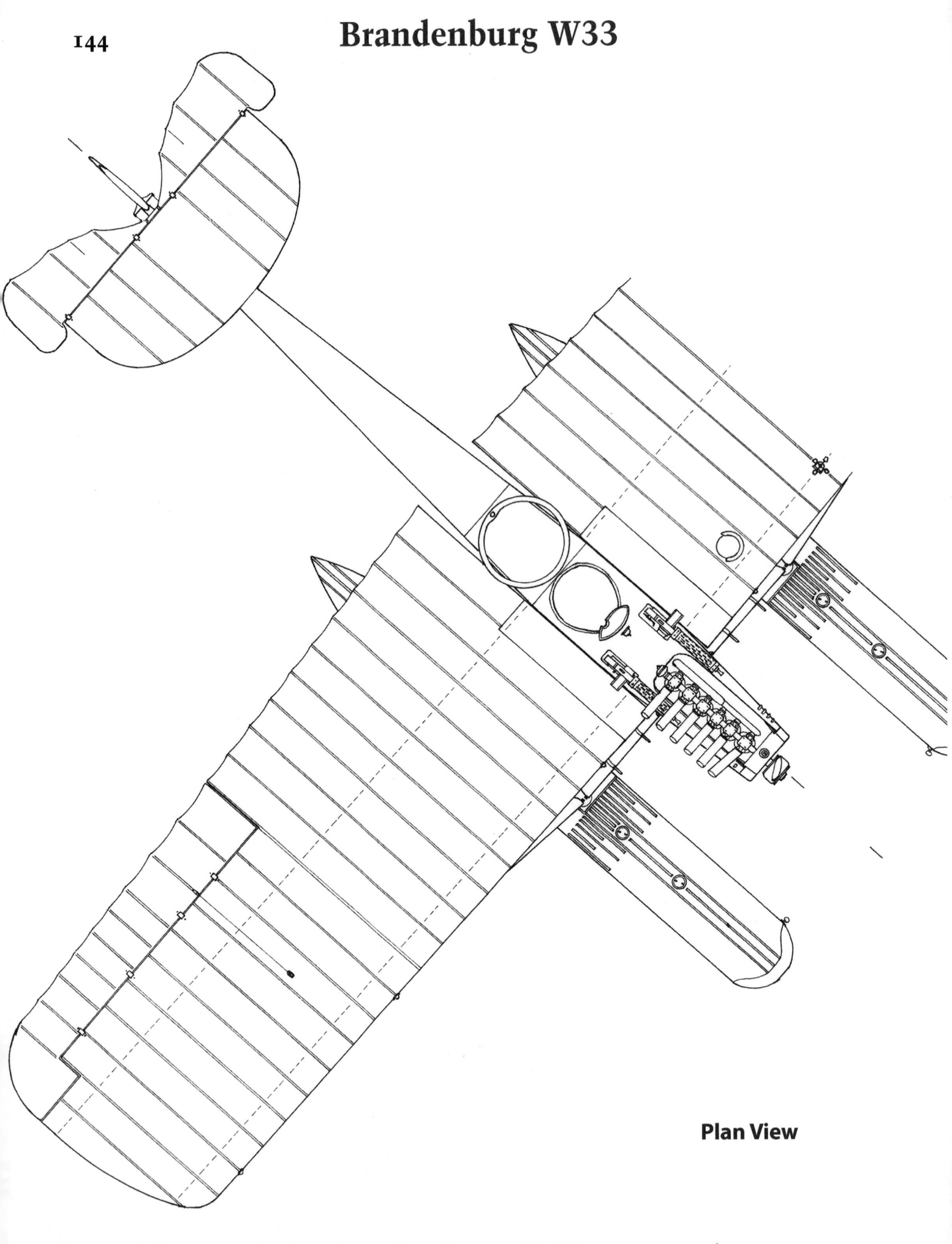

Plan View

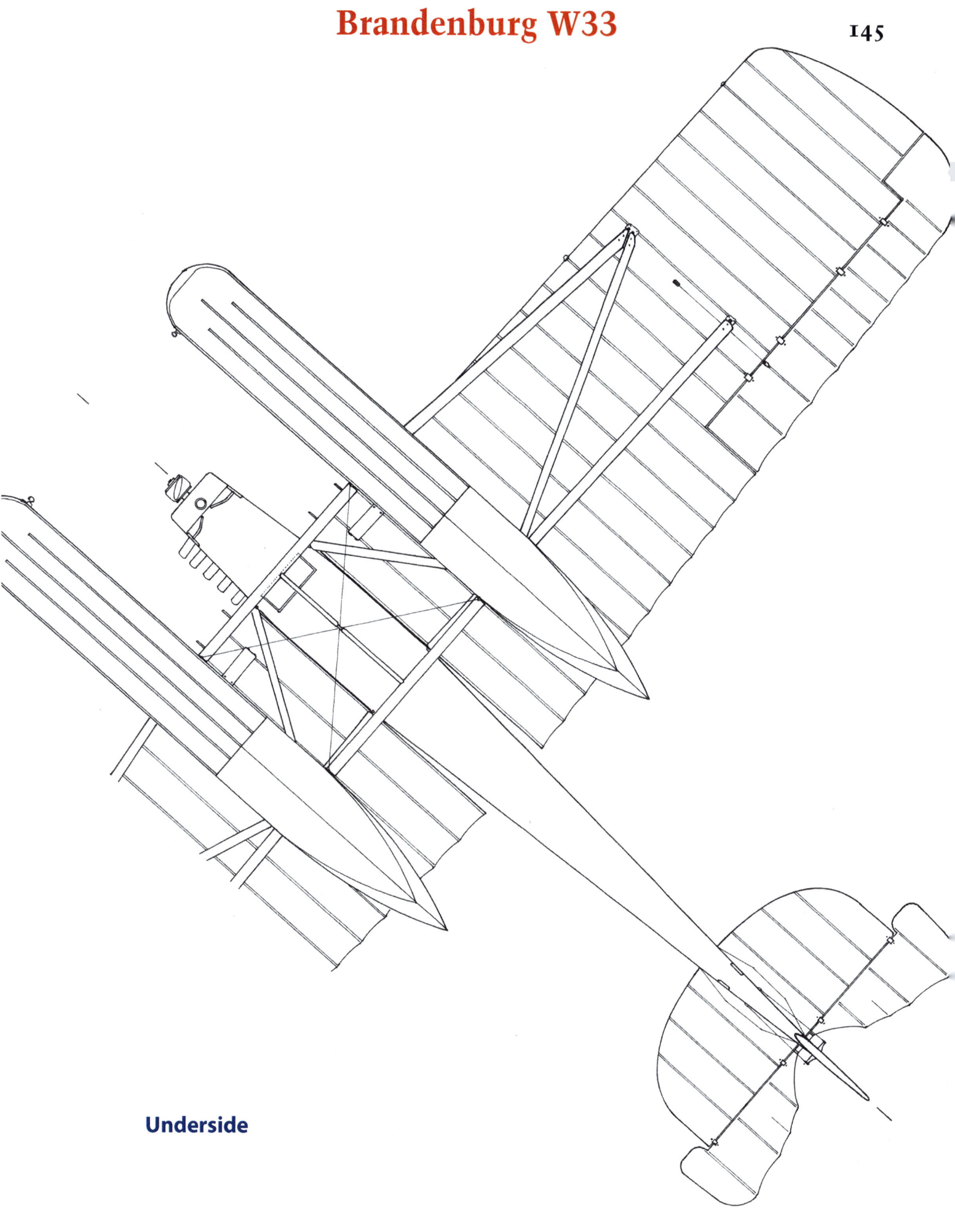

Underside

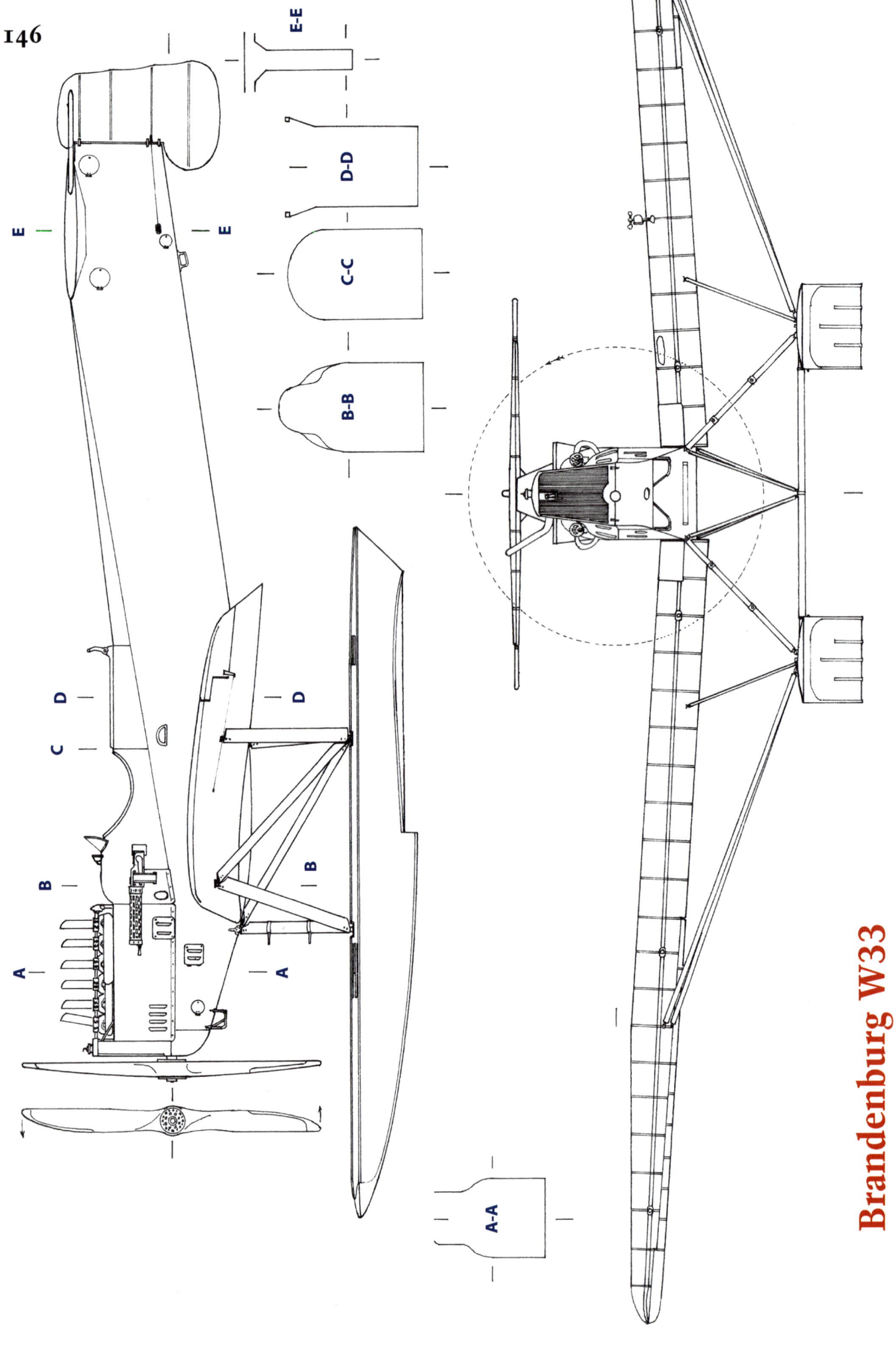

E-E

D-D

C-C

B-B

E

E

C D

D

B

B

A

A

A-A

Brandenburg W33

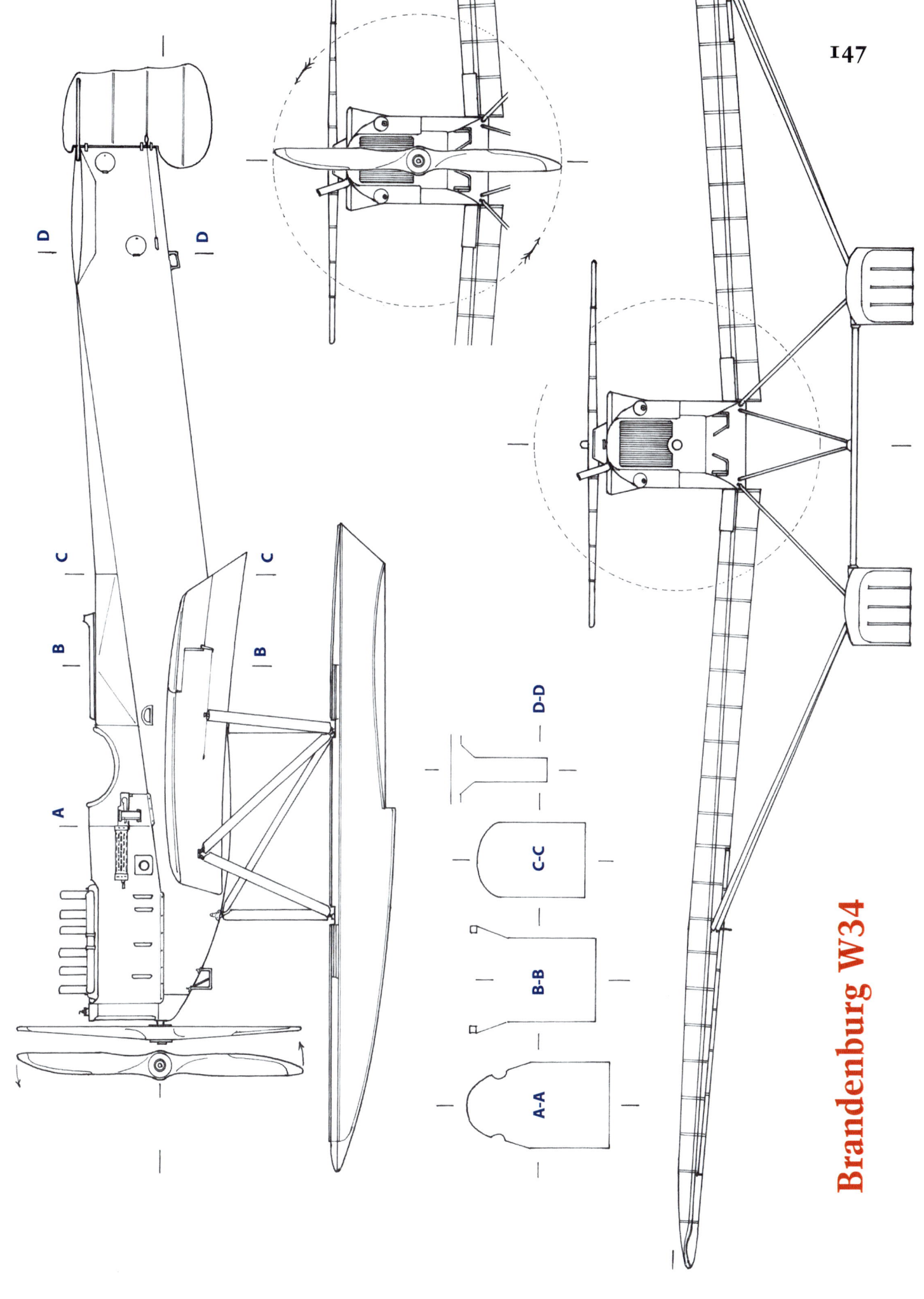

D | D

C | C

B | B

A

D-D

C-C

B-B

A-A

Brandenburg W34

Brandenburg W34

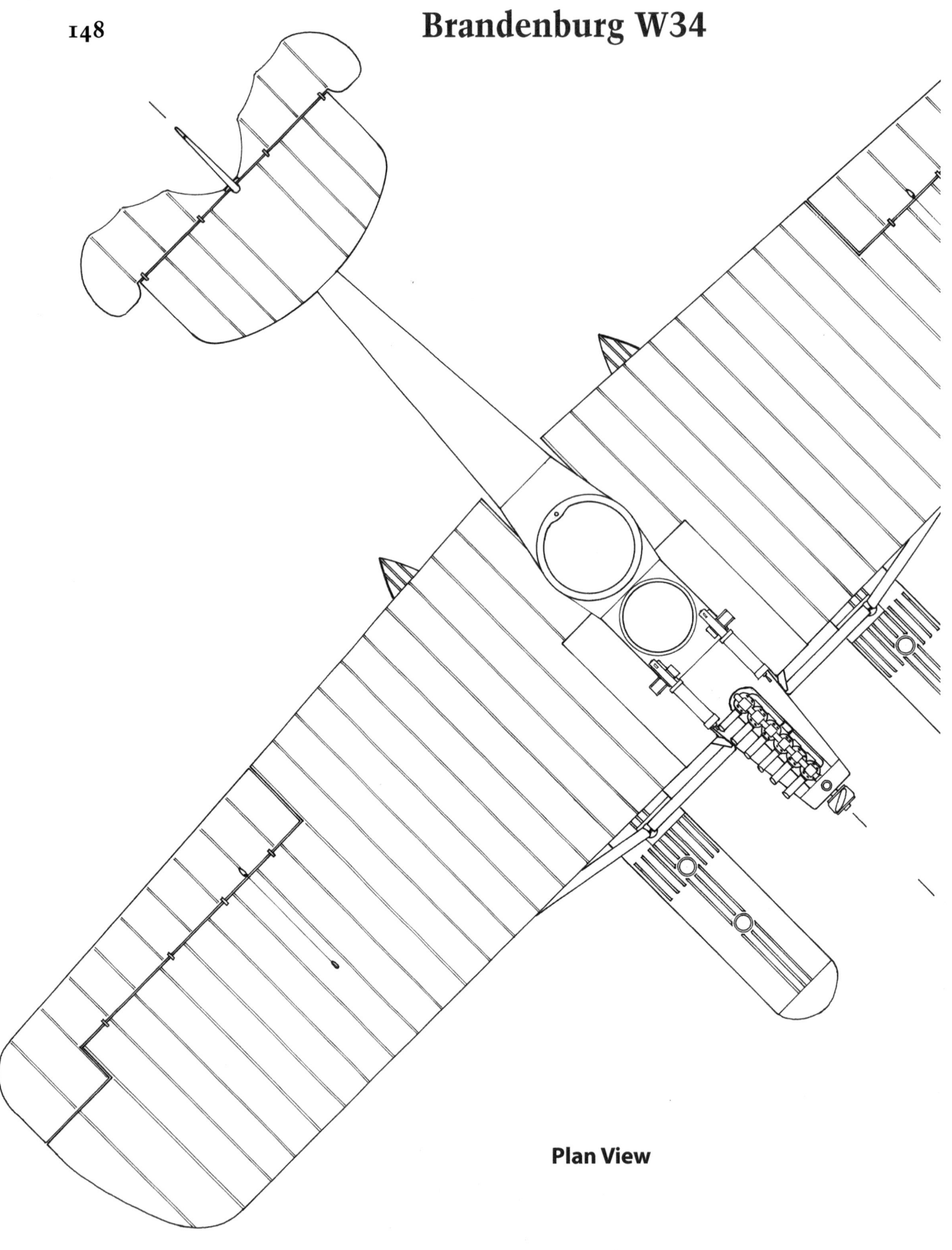

Plan View

Brandenburg W34

Underside

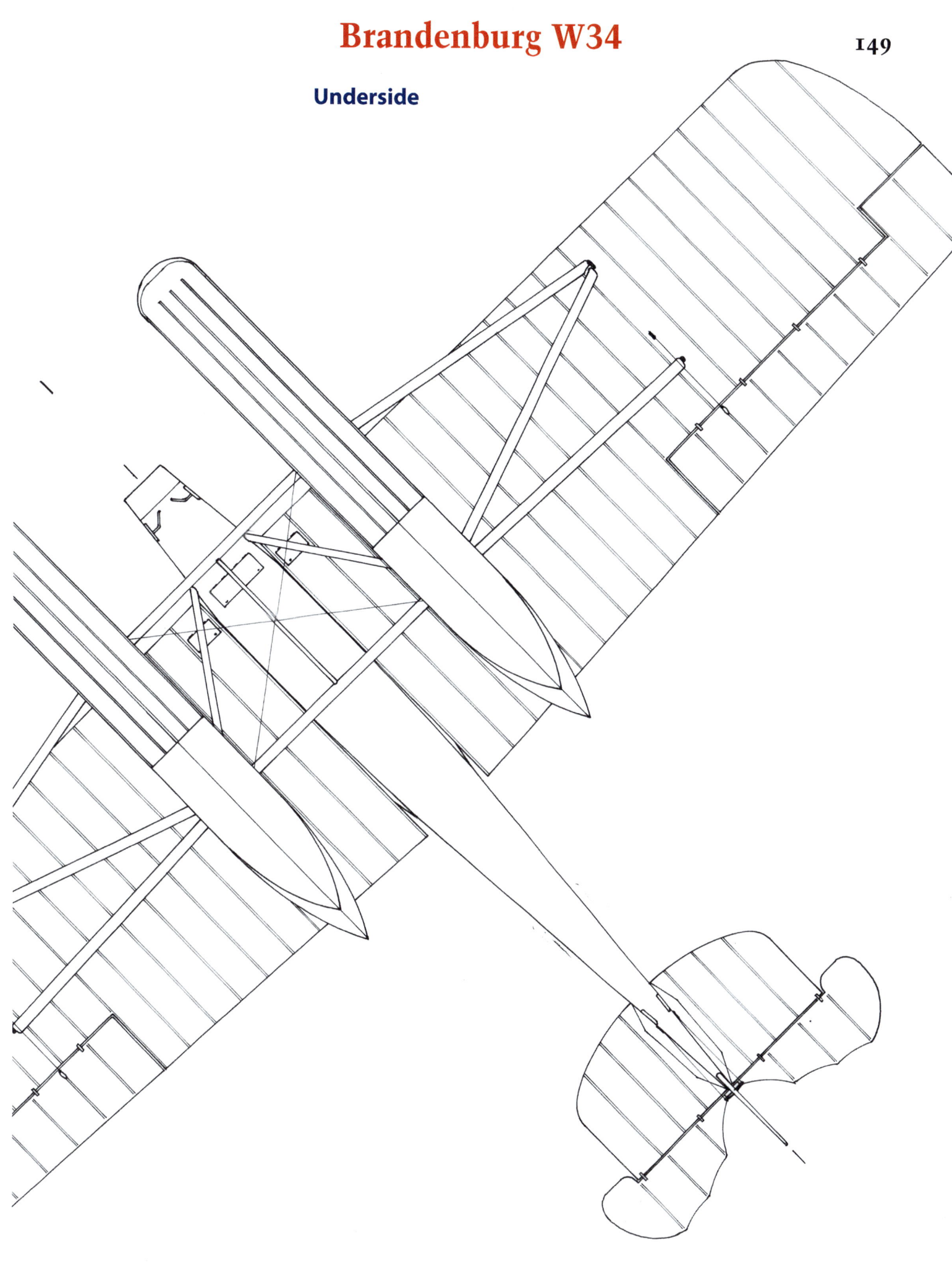

Above: An early Brandenburg KDW (no vertical fin above the fuselage) is launched as the pilot enters the cockpit. The engine appears to be a Benz due to the exhaust stack.

Below: A recently-discovered photograph of the Lübeck-Travemünde F3 single-seat fighter, Marine #844. Powered by a 150 hp Benz Bz.III, the *SVK* refused to accept it. Only one F3 was built and no other technical data has survived.

Printed in Great Britain
by Amazon.co.uk, Ltd.,
Marston Gate.